AF839799

Monitoring Ammonia-Oxidizing Bacteria in Chloraminated Distribution Systems

Awwa Research Foundation

About the Awwa Research Foundation

The Awwa Research Foundation (AwwaRF) is a member-supported, international, nonprofit organization that sponsors research to enable water utilities, public health agencies, and other professionals to provide safe and affordable drinking water to consumers.

The Foundation's mission is to advance the science of water to improve the quality of life. To achieve this mission, the Foundation sponsors studies on all aspects of drinking water, including supply and resources, treatment, monitoring and analysis, distribution, management, and health effects. Funding for research is provided primarily by subscription payments from approximately 1,000 utilities, consulting firms, and manufacturers in North America and abroad. Additional funding comes from collaborative partnerships with other national and international organizations, allowing for resources to be leveraged, expertise to be shared, and broad-based knowledge to be developed and disseminated. Government funding serves as a third source of research dollars.

From its headquarters in Denver, Colorado, the Foundation's staff directs and supports the efforts of more than 800 volunteers who serve on the board of trustees and various committees. These volunteers represent many facets of the water industry, and contribute their expertise to select and monitor research studies that benefit the entire drinking water community.

The results of research are disseminated through a number of channels, including reports, the Web site, conferences, and periodicals.

For subscribers, the Foundation serves as a cooperative program in which water suppliers unite to pool their resources. By applying Foundation research findings, these water suppliers can save substantial costs and stay on the leading edge of drinking water science and technology. Since its inception, AwwaRF has supplied the water community with more than $300 million in applied research.

More information about the Foundation and how to become a subscriber is available on the Web at **www.awwarf.org.**

Monitoring Ammonia-Oxidizing Bacteria in Chloraminated Distribution Systems

Prepared by:
John M. Regan, **Ah-Young Cho**, and **Sunhyung Kim**
Department of Civil and Environmental Engineering
The Pennsylvania State University
212 Sackett Building, University Park, Pennsylvania 16802
and
Charlotte D. Smith
Charlotte Smith & Associates, Inc.
PO Box 629, Orinda, California 94563

Sponsored by:
Awwa Research Foundation
6666 West Quincy Avenue, Denver, CO 80235-3098

Published by:

Awwa Research Foundation

American Water Works Association

IWA Publishing

DISCLAIMER

This study was funded by the Awwa Research Foundation (AwwaRF). AwwaRF assumes no responsibility for the content of the research study reported in this publication or for the opinions or statements of fact expressed in the report. The mention of trade names for commercial products does not represent or imply the approval or endorsement of AwwaRF. This report is presented solely for informational purposes.

CONTENTS

TABLES

FIGURES

FOREWORD

The Awwa Research Foundation is a nonprofit corporation that is dedicated to the implementation of a research effort to help utilities respond to regulatory requirements and traditional high-priority concerns of the industry. The research agenda is developed through a process of consultation with subscribers and drinking water professionals. Under the umbrella of a Strategic Research Plan, the Research Advisory Council prioritizes the suggested projects forwarded to the Board of Trustees for final selection. The foundation also sponsors research projects through the unsolicited proposal process; the Collaborative Research, Research Applications, and Tailored Collaboration programs; and various joint research efforts with organizations such as the U.S. Environmental Protection Agency, the U.S. Bureau of Reclamation, and the Association of California Water Agencies.

This publication is a result of one of these sponsored studies, and it is hoped that its findings will be applied in communities throughout the world. The following report serves not only as a means of communicating the results of the water industry's centralized research program but also as a tool to enlist the further support of the nonmember utilities and individuals.

Projects are managed closely from their inception to the final report by the foundation's staff and large cadre of volunteers who willingly contribute their time and expertise. The foundation serves a planning and management function and awards contracts to other institutions such as water utilities, universities, and engineering firms. The funding for this research effort comes primarily from the Subscription Program, through which water utilities subscribe to the research program and make an annual payment proportionate to the volume of water they deliver and consultants and manufacturers subscribe based on their annual billings. The program offers a cost-effective and fair method for funding research in the public interest.

A broad spectrum of water supply issues is addressed by the foundation's research agenda: resources, treatment and operations, distribution and storage, water quality and analysis, toxicology, economics, and management. The ultimate purpose of the coordinated effort is to assist water suppliers to provide the highest possible quality of water economically and reliably. The true benefits are realized when the results are implemented at the utility level. The foundation's trustees are pleased to offer this publication as a contribution toward that end.

Walter J. Bishop
Chair, Board of Trustees
Awwa Research Foundation

Robert C. Renner, P.E.
Executive Director
Awwa Research Foundation

ACKNOWLEDGMENTS

The authors of this report greatly appreciate the following water utilities and individuals for their cooperation and participation in this project:

Alameda County Water District, Fremont, Calif., Doug Chun, Cris Pena, Karl Stinson, and Karl Voigt
Ann Arbor Water Utilities, Ann Arbor, Mich., Larry Sanford and Janice Skadsen
DC Water and Sewer Authority, Washington, DC, Maureen Donnelly and Rich Giani
Louisville Water Company, Louisville, Ky., Rengao Song
Pinellas County Utilities, Clearwater, Fla., Robert Powell and Marsha Pryor
San Francisco Public Utilities Commission, San Francisco, Calif., Manouchehr Boozarpour, Michael Casteel, Andrzej Wilczak, and Alan Wong

In addition, the authors appreciated the input of the Project Advisory Committee (PAC) – including Rebecca Hoffman, Wisconsin State Lab of Hygiene; Chris Kinner, Irvine Ranch Water District; and Duncan Veal, Macquarie University – and the help of AwwaRF project officer Jennifer Warner.

EXECUTIVE SUMMARY

RESEARCH OBJECTIVES AND APPROACH

Nitrification is a common operational problem experienced by 30 to 63% of utilities that chloraminate (Wilczak et al. 1996; Seidel et al. 2005). The addition of ammonia to the distribution system associated with chloramine formation and decay can support the growth of ammonia-oxidizing bacteria (AOB). These bacteria catalyze the first step in nitrification, oxidizing ammonia to nitrite. The growth of nitrifiers is typically accompanied by a deterioration in water quality parameters such as an increase in heterotrophic plate count (HPC) bacteria, nitrite, and nitrate and a decrease in total chlorine residual (Odell et al. 1996). Once a nitrification episode is underway, increasing chloramine dose is an ineffective means of correcting the problem, requiring utilities to implement alternative control strategies (Odell et al. 1996). Therefore, it would be advantageous to detect and respond to a nitrification episode at the initial stages. The traditional culture-based method for quantifying AOB entails a three- to four-week incubation period due to their slow growth rate, so nitrification monitoring has relied on surrogate parameters of AOB growth such as nitrite and nitrate production, loss of chloramine residual, and an increase in HPC. All of these surrogates can be influenced by factors other than nitrification, and detectable changes in their concentrations can lag behind increases in the AOB population.

The overall goal of this project involved developing a reliable rapid method for the detection of AOB, allowing the monitoring of AOB numbers during nitrification episodes and responsive measures, and potentially providing chloraminating utilities an additional tool to improve their chloramination practices. The objectives of this project were to refine molecular tools for the culture-independent quantification of AOB in water samples, and to evaluate the hypothesis that an increase in AOB numbers in chloraminated systems indicates the pending onset of a nitrification episode, prior to observing changes in other water quality parameters typically associated with nitrification. The specific objectives and approach were:

(1) To expand molecular techniques for analyzing AOB populations by incorporating the ability to rapidly determine AOB numbers. Two protocols were adapted:

   a) AOB-targeted fluorescent in situ hybridization (FISH) followed by flow cytometric (FCM) analysis, and

   b) Real-time polymerase chain reaction (PCR)

(2) To monitor AOB numbers and other relevant water quality parameters during the development of nitrification episodes in several chloraminated systems, allowing the evaluation of whether an increase in AOB numbers is indicative of the pending onset of a nitrification episode. In addition to molecular AOB enumeration, this task also involved AOB measurements using the most-probable number (MPN) method, which has been applied in previous studies to characterize AOB abundance. Also, systems were monitored during the typical responsive measures of breakpoint chlorination and flushing, as well as

booster chloramination. These data reveal the responses of the AOB community to nitrification control strategies.

MAJOR FINDINGS

The most significant findings of this study are summarized as follows, organized according to the objectives listed above:

(1)(a) Adapt FISH-FCM protocol:

In FISH experiments with pure cultures, labeling AOB with a single probe did not provide sufficient fluorescent intensity to reliably discriminate AOB from non-target cells. Dual labeling with two probes (Nso1225 and AOB770) successfully increased AOB cell fluorescence relative to single-probe results, and further increases in FISH-derived AOB fluorescence were achieved using the tyramide signal amplification (TSA) approach. In adapting the microscopy-based protocols for rapid FCM analysis, a significant apparent loss of AOB was observed during successive filtration and resuspension steps, with recoveries of 60-82% for *Nitrosomonas europaea* and 7-14% for *Nitrosospira multiformis*.

(1)(b) Adapt real-time PCR protocol:

The real-time PCR assay targeting the AOB-specific functional gene ammonia monooxygenase (*amoA*) provided excellent sensitivity and linearity, allowing quantitation down to just over 100 *amoA* copies. With the filtration of 1-L samples and correction for dilution of DNA in the PCR mixture, this offered detection limits between 4 to 6 *amoA* copies/mL. This detection limit could be lowered by filtering larger sample volumes. Based on recoveries from spiked samples, the assay did not show appreciable PCR inhibition from any of the water matrices.

(2) Monitor full-scale chloraminated systems:

- The results from a reservoir sampled in this study showed a good correspondence between AOB numbers (determined by real-time PCR) and changes in chemical indicators of nitrification. The nitrite and AOB data generally followed the same trend, with elevated nitrite concentrations coinciding with detectable AOB concentrations relative to a non-detect background. A reduction in chloramine residual also generally matched the AOB and nitrite changes, and no AOB were detected at chloramine concentrations above 2 mg/L in this system. (Several of the systems included in the study did not show evidence of nitrification over the course of the sampling events.)
- In two closed-loop piping systems monitored in this study, there was an apparent disconnect between suspended AOB numbers and changes in water quality parameters impacted by nitrification. This may indicate that the nitrifying community in these systems included a significant biofilm component not represented in the grab samples apart from detached cells.
- Based on the collective findings from the systems that showed evidence of nitrification, AOB numbers from liquid grab samples do not serve as an early indication of a nitrification

episode. While the reservoir system generally showed concurrent changes in biological and chemical water quality, the two pipe loop systems did not consistently demonstrate this trend.

- The real-time PCR protocol showed a significant improvement over the MPN cultivation approach. Not only did the method yield results within one day compared to a three to four week incubation period for MPN, but it also showed dramatically higher AOB numbers.
- Flushing of one of the pipe-loop systems in response to loss of chloramine residual and increase in nitrite was not an effective remedy for nitrification control. Evidence of nitrification returned each time in less than one week. This may be related to a biofilm-influenced AOB community.
- Based on the several-month period during which it was monitored in this study, booster chloramination in a storage reservoir appeared to be an effective nitrification control strategy. Maintaining the chloramine level above 2 mg/L prevented AOB growth, and only during the period in which the booster system was turned off did chloramine levels drop and nitrite and AOB concentrations increase.
- Real-time PCR results, confirmed by cloning and sequencing of *amoA* fragments, showed the exclusive detection of *Nm. oligotropha*-like AOB in the chloraminated systems monitored in this study.

RECOMMENDATIONS

The following recommendations are made based on the results of this study:

1. The use of FCM for rapid quantitation of AOB would require protocol improvements to address AOB losses associated with cell filtration and resuspension, as well as modifications to reduce the volume of reagents and the associated protocol costs. TSA-FISH was effective at boosting fluorescent signals of AOB, and coupled with FCM might provide a rapid analysis method. This might also present an effective strategy for the molecular enumeration of other microbial groups in drinking water.
2. The real-time PCR protocol used in this study presents a marked improvement over MPN enumeration, with respect to both turnaround time and cell recovery. Systems that are presently complementing their chemical analyses with MPN data might consider adapting this molecular approach.
3. The results of the pipe-loop systems, in contrast to those from a reservoir that experienced nitrification during this study, may suggest a significant role of biofilm-associated AOB in these episodes. In one of the systems, repeated flushing did not remedy the nitrification problem. In the other system, there were clear chemical indications of nitrification, but AOB were below the detection limit for over two weeks. The methods used in this study focused solely on liquid grab samples, thereby missing this biofilm contribution. Future research should investigate the role of biofilms at harboring AOB, and directly assess the effectiveness of nitrification control strategies at inactivating AOB in biofilms.
4. Booster chloramination may be an effective approach for controlling nitrification in reservoirs. While this study only monitored booster chloramination for a few months, it appeared to eradicate AOB from suspension and eliminated the associated production of nitrite.

CHAPTER 1
INTRODUCTION

OVERVIEW

Many drinking water utilities in the United States have converted from chlorine to chloramines for secondary disinfection (AWWA 2000; Seidel et al. 2005). The current use of chloramines is approximately 30% of utilities, and this is projected to show an increase in response to increasingly stringent regulations aimed at controlling disinfection byproducts (Seidel et al. 2005). Chloramination has been used successfully to control the formation of trihalomethanes (THMs) and haloacetic acids (HAAs) (Brodtmann, and Russo 1979; Norman, Harms, and Looyenga 1980; Norton, and LeChevallier 1997; Carlson, and Hardy 1998).

Nitrification is a common operational problem experienced by 30 to 63% of utilities that chloraminate (Wilczak et al. 1996; Seidel et al. 2005). The addition of ammonia to the distribution system associated with chloramine production during the treatment process and chloramine decay within the distribution system can support the growth of ammonia-oxidizing bacteria (AOB). These bacteria catalyze the first step in nitrification, oxidizing ammonia to nitrite. The growth of nitrifiers is typically accompanied by a deterioration in water quality parameters such as an increase in heterotrophic plate count (HPC) bacteria, nitrite, and nitrate and a decrease in total chlorine residual (Odell et al. 1996). Once a nitrification episode is underway, increasing chloramine dose is an ineffective means of correcting the problem, requiring utilities to implement alternative control strategies (Odell et al. 1996). Therefore, it would be advantageous to detect and respond to a nitrification episode at the initial stages.

The prohibitively slow growth rate of AOB precludes the use of culturing methods such as the most-probable number (MPN) technique (Matulewich, Strom, and Finstein 1975; Belser, and Mays 1982) to gain operationally meaningful enumeration data. Therefore, the monitoring of nitrification in these systems is based on surrogate parameters such as nitrite, total chlorine residual, and HPC bacteria (Wolfe et al. 1990; Cunliffe 1991; Lieu, Wolfe, and Means 1993; Wilczak et al. 1996). However, these efforts have failed to demonstrate a consistent correlation between these potential surrogate indicators and nitrification episodes for predictive purposes. Wolfe *et al.* (1990) generally observed poor correlations between AOB numbers and nitrite or chlorine concentrations, and an increase in nitrite concentration has been reported to lag behind AOB growth (Ike, Wolfe, and Means 1988; Lieu, Wolfe, and Means 1993). One complicating factor is that non-nitrifying conditions can also influence the symptoms of nitrification, making the very definition of nitrification episodes imprecise based on these associated indicators. The ability to rapidly detect and enumerate AOB would provide an additional monitoring tool for the nitrification control strategies of chloraminating utilities.

Recent advances in the development of quantitative, culture-independent molecular detection tools provide a new and promising opportunity to explore the direct use of AOB numbers, instead of only surrogate parameters, for monitoring nitrification in chloraminated systems. There has been considerable application of molecular techniques to the study of nitrifying bacteria due to their slow growth rates and the failure of culturing to represent the true diversity of nitrifier communities. The majority of this research has focused on the development and application of nucleic acid-based techniques for the specific detection and characterization of nitrifiers in soil, sediment, wastewater, and water environments. These include recent studies of nitrifying communities in chloraminated distribution systems (Regan, Harrington, and Noguera 2001; Regan, Harrington, and Noguera 2002), which found the exclusive presence of AOB related to *Nitrosomonas oligotropha*, an AOB species that until recently had been largely overlooked in freshwater environments. While these studies provided a qualitative characterization of AOB that had become established during nitrification episodes in a pilot-scale and several full-scale distribution systems, an attempt to *quantify* AOB in these systems and monitor the establishment and proliferation of an AOB community during system operation has not yet been reported.

Two molecular techniques that have often been used to study nitrifying bacteria include fluorescent in situ hybridization (FISH) (Wagner et al. 1995; Mobarry et al. 1996; Wagner et al. 1996; Gieseke et al. 2001) and the polymerase chain reaction (PCR) (Sinigalliano, Kuhn, and Jones 1995; Stephen et al. 1996; Rotthauwe, Witzel, and Liesack 1997). FISH is a whole-cell technique that involves the specific hybridization of a fluorescently labeled probe to the ribosomal RNA (rRNA) of target organisms (Amann, Krumholz, and Stahl 1990). rRNA is an abundant molecule in active cells, potentially allowing sufficient probe-conferred fluorescence within target cells to allow their discrimination from non-target organisms. FISH has been extensively used to study AOB communities in a variety of environments, including laboratory-scale reactors and environmental samples (Wagner et al. 1995; Mobarry et al. 1996; Pommerening-Röser, Rath, and Koops 1996; Schramm et al. 1996; Wagner et al. 1996; Juretschko et al. 1998; Schramm et al. 1998; Schramm et al. 1999; Okabe, and Watanabe 2000; Gieseke et al. 2001). While this technique enables the collection of quantitative data, its typical implementation with microscopic visualization is tedious and time consuming, making it impractical in this form for routine system monitoring.

PCR is an enzymatic technique for generating numerous copies of a target fragment of DNA, which can subsequently be detected by a variety of methods. It is a cyclic procedure that theoretically involves a doubling of amplification products with each cycle. Traditional PCR-based protocols are at best semi-quantitative, but there are several methodological issues that can introduce quantitation errors in PCR-generated data (Wintzingerode, Göbel, and Stackebrandt 1997). Recently PCR techniques have been developed that provide quantitative data, including a method called real-time PCR that accomplishes a precise quantitative measure of specific target sequences from the initial exponential phase of amplification (Heid et al. 1996). Real-time PCR has been applied to the quantification of AOB in several environments, including soil (Hermansson, and Lindgren 2001) and wastewater (Layton et al. 2005). However, it has not been previously applied to study AOB numbers in chloraminated systems.

OBJECTIVES AND APPROACH

The overall goal of this project involved the rapid detection of AOB, allowing the monitoring of AOB numbers during nitrification episodes and responsive measures, and potentially providing chloraminating utilities an additional tool to improve their chloramination practices. The objectives of this project were to refine molecular tools for the culture-independent quantification of AOB in water samples, and to evaluate the hypothesis that an increase in AOB numbers in chloraminated systems indicates the pending onset of a nitrification episode, prior to observing changes in other water quality parameters typically associated with nitrification. The specific objectives and approach were:

(1) To expand molecular techniques for analyzing AOB populations by incorporating the ability to rapidly determine AOB numbers. Two protocols were adapted:

- AOB-targeted FISH followed by flow cytometric analysis (Amann et al. 1990; Wallner, Erhart, and Amann 1995; Fuchs et al. 1998) for rapid quantitation. Protocol enhancements, including hybridization with multiple probes and tyramide signal amplification (TSA)(Biegala et al. 2003), were explored to increase the fluorescent intensities of AOB, with the goal of allowing time-savings associated with flow cytometric enumeration rather than microscopic analysis.

- Real-time PCR. A real-time PCR protocol was recently developed for the rapid quantification of AOB in soil (Hermansson, and Lindgren 2001). While the approach was successful, the primers that were selected for the protocol are known to have several nucleotide mismatches with the 16S rRNA gene (16S rDNA) of certain AOB groups. This includes a three nucleotide mismatch with the 16S rDNA sequences of *Nitrosomonas oligotropha* representatives (Purkhold et al. 2000), an AOB cluster that has been shown to be a significant contributor to nitrification episodes in chloraminated systems (Regan, Harrington, and Noguera 2001; Regan, Harrington, and Noguera 2002). Hence, it is likely that the existing PCR protocol would generate false negatives, which would jeopardize the success of this study.

(2) To monitor AOB numbers and other relevant water quality parameters during the development of nitrification episodes in several chloraminated systems, allowing the evaluation of whether an increase in AOB numbers is indicative of the pending onset of a nitrification episode. In addition to molecular AOB enumeration, this task also involved AOB measurements using the MPN method, which has been applied in previous studies to characterize AOB abundance. Also, systems were monitored during the typical responsive measures of breakpoint chlorination and flushing, as well as booster chloramination. These data reveal the responses of the AOB community to nitrification control strategies.

PROJECT SIGNIFICANCE

This project fills a data gap that exists in previous studies, providing information on the direct impact of chloramination on nitrifier growth to complement the data on the associated evidence or symptoms of nitrifier growth. This research also contributes a characterization of the progression of nitrification episodes, whereas much of the published literature on this topic offers only limited snapshots of evidence of nitrification in chloraminated systems. The method described in this report offers a rapid and sensitive technique for quantifying AOB concentrations in drinking water samples, providing a marked improvement over the culture-based technique presently used by the water industry by reducing analysis turnaround time from at least three weeks to one or two days and allowing the quantitation of environmental strains that prove difficult to recover by culturing.

The molecular measurement protocols could be incorporated in a distribution system monitoring plan, with action levels determined by an observed growth of AOB. Presently, many chloraminated systems have an action level based on a threshold concentration of nitrite or nitrate. However, a detectable increase in nitrite concentration has been shown to lag behind

AOB growth (Ike, Wolfe, and Means 1988; Lieu, Wolfe, and Means 1993). The reliance on nitrite alone as an indicator delays the response to the onset of nitrification, potentially making nitrification control more difficult and introducing the possibility of water quality impacts. Responding to the initial evidence of nitrifier growth could limit the extent and impact of nitrification episodes in chloraminated systems.

CHAPTER 2
BACKGROUND

CHLORAMINATION

Secondary Disinfection with Chloramines

Chlorination has been the most common method of disinfection in the United States since the early 1900s. Chlorine is very economical and effective in killing waterborne pathogenic bacteria and viruses and is therefore appropriate as a primary disinfectant. However, when free chlorine is added to water containing natural organic matter (NOM), these species react to form undesirable chlorinated organic byproducts such as trihalomethanes (THMs) and haloacetic acids (HAAs) (Norman, Harms, and Looyenga 1980). In addition, brominated byproducts can be formed when source water containing bromide ions is chlorinated. Chlorine oxidizes bromide to free bromine, which can react with NOM to produce the brominated analogs of chlorinated byproducts (Krasner et al. 1989). The US Environmental Protection Agency (USEPA) promulgated the Disinfectants/Disinfection By-Products (D/DBP) rule (USEPA 1998) to establish standards for disinfection byproducts caused by chlorination and have continued to lower allowable concentrations since, through the promulgation of additional rules.

To comply with these rules, many drinking water utilities have switched to the use of chloramines as an alternative secondary disinfection method (Carlson, and Hardy 1998). According to a recent report by the American Water Works Association, chloramine use at medium to large facilities has increased from 20 percent in 1989 to 29 percent in 1998 (AWWA 2000). A more recent survey showed a similar chloramine usage, with a projection for a slight increase in the coming years (Seidel et al. 2005). Much of this switch to chloramines has been prompted by the increasingly stringent regulations on chlorinated disinfection byproducts and the need to reduce biological regrowth (Wilczak et al. 1996). Chloramines have been shown to produce lower concentrations of total THMs and HAAs than free chlorine (Brodtmann, and Russo 1979; Norman, Harms, and Looyenga 1980), with reported THM reductions from 40 to 80 percent (Kirmeyer et al. 1993). There are also data from full-scale demonstrations that indicate a potential improvement in residual maintenance and regrowth control offered by chloramines (Neden et al. 1992; Norton, and LeChevallier 1997) because chloramine residuals are more stable in distribution systems than free chlorine residuals. Chloramines can also help to reduce tastes and odors, especially from phenols (Norton, and LeChevallier 1997).

Although chloramination significantly reduces the levels of regulated chlorinated byproducts, several potential problems may arise. Chloramines produce higher levels of cyanogen chloride than free chlorine. The conversion of cyanogen chloride to cyanide and thiocyanate could contribute to the chronic toxicity of chloraminated drinking water (Simpson, and Hayes 1998). Recently, N-nitrosodimethylamine (NDMA), which is considered a probable human carcinogen, was also detected in some California water utilities. NDMA formation is directly related to the concentration of chloramines and the contact time with free chlorine (Wilczak et al. 2003). In addition to these chemical problems, a microbiological concern regarding the use of chloramines is the potential for nitrification. The relationship between chloramination and nitrification, and the problems associated with nitrification, are discussed below.

Chlorine and Ammonia Chemistry

Chlorine is generally applied in drinking water disinfection as molecular chlorine ($Cl_{2(g)}$ or $Cl_{2(aq)}$), sodium hypochlorite (NaOCl), or calcium hypochlorite ($Ca(OCl)_2$) . When chlorine gas is added to water, it rapidly partitions into gas and aqueous phases according to Henry's Law (Haas 1990) and undergoes the following disproportionation reaction:

$$Cl_{2(aq)} + H_2O \leftrightarrow HOCl + H^+ + Cl^- \qquad (2.1)$$

Hypochlorous acid (HOCl), a weak acid with a pK_a of 7.5 at 25°C (Snoeyink, and Jenkins 1980), is in equilibrium with hypochlorite (OCl^-) according to the following:

$$HOCl \leftrightarrow H^+ + OCl^- \qquad (2.2)$$

Each of the hypochlorite salts dissociates to hypochlorite and the corresponding cation:

$$NaOCl \rightarrow Na^+ + OCl^- \qquad (2.3)$$

$$Ca(OCl)_2 \rightarrow Ca^{2+} + 2\,OCl^- \qquad (2.4)$$

The sum of $Cl_{2(aq)}$, HOCl, and OCl^- is referred to as free available chlorine, with the relative composition depending on pH. Only trace molecular chlorine is present at pH values relevant in drinking water. Hypochlorous acid predominates below pH 7.5 and is typically two orders of magnitude more effective as a disinfectant than hypochlorite ion (White 1999).

Chloramines are formed by the reaction of hypochlorous acid and ammonia, producing a series of chlorinated ammonia compounds. These are referred to as combined chlorine, and the sum of free and combined chlorine is called total chlorine. The following reactions characterize the formation of the three inorganic chloramine species:

$$NH_3 + HOCl \leftrightarrow NH_2Cl + H_2O \text{ (monochloramine)} \qquad (2.5)$$

$$NH_2Cl + HOCl \leftrightarrow NHCl_2 + H_2O \text{ (dichloramine)} \qquad (2.6)$$

$$NHCl_2 + HOCl \leftrightarrow NCl_3 + H_2O \text{ (trichloramine or nitrogen trichloride)} \qquad (2.7)$$

Monochloramine is the predominant form at pH values greater than 7 and a chlorine to ammonia-nitrogen weight ratio less than 5 to 1 (mole ratio less than 1 to 1). Dichloramine, which causes taste and odor problems but is a more effective disinfectant than monochloramine (Oldenburg et al. 2002), becomes significant as pH is lowered and at a higher chlorine to ammonia-nitrogen ratio. Trichloramine is a minor constituent of chloramines until the pH is below 4 (Snoeyink, and Jenkins 1980). As the chlorine dose increases for a given ammonia concentration and the chlorine to ammonia-nitrogen mass ratio increases above 5:1, the chlorine residual decreases due to the oxidation of ammonia to nitrogen gas. The total chlorine residual reaches a minimum level close to zero at the breakpoint. Prior to the breakpoint, combined chlorine as monochloramine and dichloramine predominates. However, after the breakpoint, HOCl and OCl^- become the dominant species since the ammonia has been removed. To control tastes and odors and to prevent the loss of chloramine residual caused by the breakpoint reaction, water utilities typically maintain a chlorine to ammonia-nitrogen weight ratio between 4:1 and 5:1 (Kreft et al. 1985).

The link between chloramination and nitrification stems from the addition of ammonia to the distribution system. This added ammonia is associated with both residual ammonia from chloramine formation and ammonia released during chloramine decay. Table 2.1 shows several chloramine decay reactions that release free ammonia in the distribution system. As discussed in the next section, it is this ammonia that potentially promotes biological instability by stimulating the growth of nitrifying bacteria.

Table 2.1
Chloramine decay reactions that release ammonia to the distribution system

Reaction	Stoichiometry	Equation
Chloramine auto-decomposition	$3\ NH_2Cl \rightarrow N_2 + NH_4^+ + 3\ Cl^- + 2\ H^+$	2.8
Oxidation of organic matter by chloramine	$0.1\ C_5H_7O_2N + NH_2Cl + 0.9\ H_2O \rightarrow$ $0.4\ CO_2 + 0.1\ HCO_3^- + 1.1\ NH_4^+ + Cl^-$	2.9
Reaction of chloramine with corrosion products at pipe wall	$0.5\ NH_2Cl + H^+ + Fe^{2+} \rightarrow$ $Fe^{3+} + 0.5\ NH^{4+} + 0.5\ Cl^-$	2.10
Oxidation of nitrite by chloramine	$NH_2Cl + NO_2^- + H_2O \rightarrow$ $NH_3 + NO_3^- + HCl$	2.11

Source: Adapted from Woolschlager et al. 2001.

NITRIFICATION IN CHLORAMINATED SYSTEMS

Occurrence and Symptoms of Nitrification

Nitrification is a common operational problem associated with chloramination. A survey of drinking water utilities using chloramines as a secondary disinfectant revealed that 63 percent of the utilities had experienced evidence of nitrification occurrence (Wilczak et al. 1996). A more recent survey showed only 30 percent of chloraminating utilities observed nitrification in their systems (Seidel et al. 2005). The residual and released ammonia associated with chloramination can stimulate nitrification, a two-step microbial process that involves the serial oxidation of ammonia to nitrite by ammonia-oxidizing bacteria (AOB) and nitrite to nitrate by nitrite-oxidizing bacteria (NOB). The energy-yielding reactions catalyzed by these distinct functional groups are shown below:

$$NH_3 + 3/2\ O_2 \rightarrow NO_2^- + H^+ + H_2O \quad (2.12)$$

$$NO_2^- + \tfrac{1}{2}\ O_2 \rightarrow NO_3^- \quad (2.13)$$

Both groups of nitrifiers are aerobic autotrophs, using oxygen as a terminal electron acceptor and carbon dioxide as a carbon source. Hence, ammonia is typically the rate-limiting growth substrate. Even if the chlorine to ammonia-nitrogen ratio is properly maintained, free ammonia is always present in a chloraminated distribution system, and this ammonia can promote nitrifier growth in otherwise stable finished water. Although NOB have been detected in chloraminated systems (Regan, Harrington, and Noguera 2002; Regan et al. 2003), incomplete nitrification is often observed in these systems by the detection of the nitrite intermediate. Therefore, the focus of this study was on AOB, which initiate the cascade of nitrification symptoms.

While nitrifying bacteria do not pose a direct threat to public health, nitrification episodes are marked by a change in several water quality parameters, some of which may have public health implications. These include a decrease in chloramine residual, dissolved oxygen (DO), pH, and alkalinity; and an increase in heterotrophic plate count (HPC) bacteria, nitrite, and nitrate (Wolfe et al. 1988; Skadsen 1993; Kirmeyer et al. 1995; Odell et al. 1996; Smith, and Smith 1996; Wilczak et al. 1996). The growth of AOB and the production of nitrite can deplete the chloramine residual that is responsible for disinfection (Equation 2.11). Nitrite is also problematic because it can further promote biological instability due to the growth of NOB, and can react with secondary amines to produce nitrosamines, which are known carcinogens (DiGiano, Carrier, and Dietrich 1986). Eventually, destruction of the chloramine residual and the growth of AOB can lead to the development of a microbial biofilm community, which could result in water quality violations under the Total Coliform Rule and the Surface Water Treatment Rule (Wilczak et al. 1996).

Figure 2.1 illustrates a conceptual model of how chloramination can give rise to a nitrification episode and the accompanying impacts on water quality. Free ammonia stimulates the growth of AOB, which in turn produce nitrite and soluble organic compounds (Rittmann, Regan, and Stahl 1994). These products of AOB growth and decay can promote the growth of NOB and heterotrophic bacteria. Moreover, nitrite and the organics associated with the

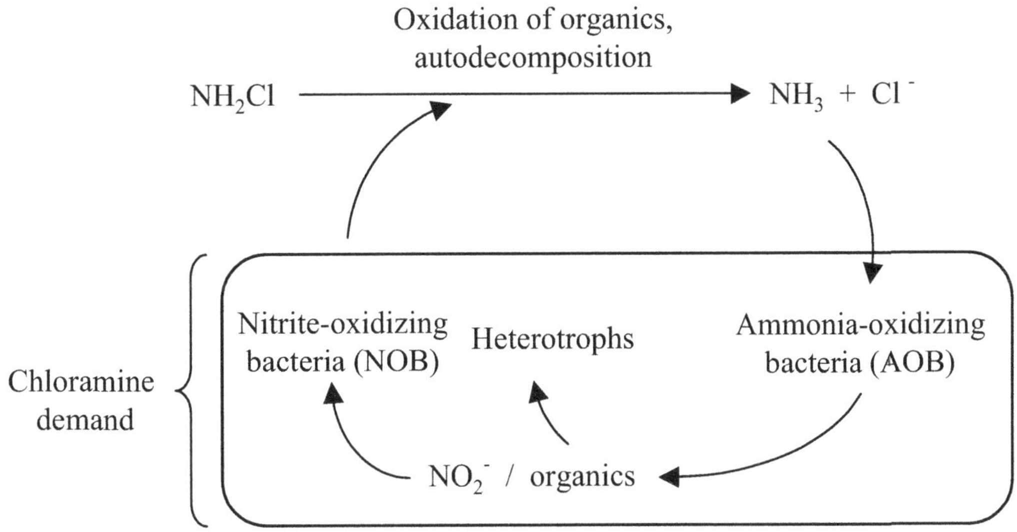

Figure 2.1 Conceptual model of chloramine-induced biological regrowth

microbial growth exert a chloramine demand. This exacerbates the problem by reducing the chloramine concentration intended to prevent microbial growth and increasing the ammonia concentration that initiates the cycle by promoting AOB growth. The net effect involves an increase in microbial growth and growth products and a decrease in residual chloramines. While it has not been proven whether nitrification episodes cause the associated loss in chloramine residual or develop in response to a decreased residual, this feedback loop is consistent with the observation that it is very difficult to control a nitrification episode by increasing chloramine dose (Skadsen 1993; Odell et al. 1996).

Controlling or Limiting Nitrification

Several approaches have been recommended for the prevention and control of nitrification. Breakpoint chlorination is often used effectively in response to an existing nitrification episode. Thus, many utilities implement breakpoint chlorination once or twice yearly for a period of one week to one month in an attempt to prevent nitrification (Kirmeyer et al. 1995). However, occasional breakpoint chlorination is not an effective long-term strategy, because it does not remedy the conditions that support nitrification. Moreover, it is unknown whether there would be a DBP-related health risk associated with periodic breakpoint chlorination for those utilities that switched to chloramination to comply with DBP limits.

The conventional wisdom, based on documented success in some systems (Wolfe et al. 1990; Norton, and LeChevallier 1997), has been to maintain a chloramine residual above 2 mg/L. This may reduce nitrification events, but once symptoms of nitrification appear, increasing the chloramine dose is not an effective remedy. Cunliffe (1991) observed evidence of nitrification in 20% of systems with residuals above 5 mg/L, and Skadsen (1993) noted that increasing the dose as high as 8 mg/L did not remedy an exisiting nitrification problem.

A reduction of the available free ammonia in the distribution system is another long-term preventive measure. This can be accomplished by increasing the dosing ratio of chlorine to ammonia-nitrogen to limit the residual ammonia from chloramine formation. The effectiveness of this strategy has not been clearly demonstrated; while chloramines appear to be more effective biocides at mass ratios above 3:1 at 25°C (Lieu, Wolfe, and Means 1993), the chlorine to ammonia ratio did not affect biocide efficacy at 10 and 15°C (Lieu, Wolfe, and Means 1993), and systems with ratios as high as 4.5:1 have shown evidence of nitrification (Skadsen 1993; Odell et al. 1996). Implementing measures to limit chloramine decay offers another approach to reducing the free ammonia concentration. Pilot studies comparing the effect of conventional and enhanced coagulation on chloramine decay and nitrification showed mixed results, but generally enhanced coagulation reduced chloramine decay and delayed the onset of nitrification (Harrington et al. 2003). A possible explanation for the growth of AOB even at the very low free ammonia concentrations characteristic of these systems is that *Nitrosomonas oligotropha*, the species that has been detected in chloraminated systems (Regan, Harrington, and Noguera 2002; Regan et al. 2003), has a half-maximum substrate concentration (K_s) about an order of magnitude lower than other *Nitrosomonas* species (Stehr et al. 1995; Gieseke et al. 2001). This means they have a significantly higher affinity for ammonia than other characterized AOB species, attaining half their maximum growth rate at one-tenth the ammonia concentration.

A recently suggested approach for preventing AOB growth couples the affects of chloramine residual and ammonia concentration (Fleming, Harrington, and Noguera 2005). The former is responsible for AOB inactivation, while the latter affects AOB growth. A nitrification potential curve can be constructed on a plot of total chlorine versus free ammonia by accounting for inactivation and growth (Figure 2.2). Using Chick-Watson kinetics to describe inactivation and Monod kinetics to model growth, this curve follows a saturation function with a maximum defined by the ratio of specific growth and inactivation rates (R_{gi}), and the half-maximum defined by K_s. Conditions above the potential curve should be less prone to nitrification. It remains to be seen whether this approach is successful in overcoming the apparent discrepancies found in predicting nitrification based solely on chloramine or ammonia concentration.

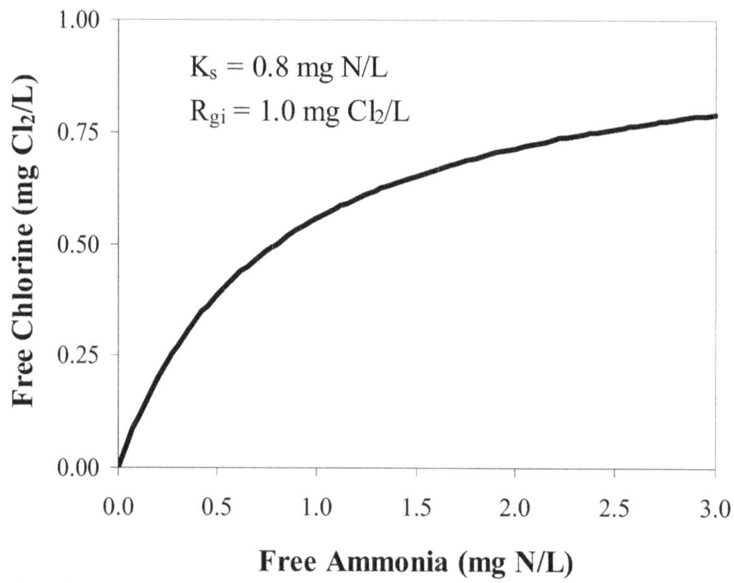

Source: Adapted from Fleming, Harrington, and Noguera 2005.

Figure 2.2 Nitrification potential curve

The pH also can be a significant factor in controlling nitrification problems. Apart from corrosion considerations, operating at a pH below 7 may prevent the growth of nitrifiers for two reasons: nitrifier activity is much lower below pH 7 or above pH 9; and a lower pH improves chloramine effectiveness and NOM removal by coagulation, both of which provide more biologically stable water (Oldenburg et al. 2002; Harrington et al. 2003). Coversely, a high pH has also been suggested as an effective nitrification control strategy. The Ann Arbor Water Utilities reported success at controlling nitrification in their system after raising their pH above 9.3 (Skadsen 2002). A survey of chloraminated systems indicated apparent nitrification over a pH range of 6.6 to 9.7 (Odell et al. 1996).

Chlorite ion (ClO_2^-) is another strategy for controlling nitrification. Chlorite has been shown to effectively inhibit AOB at levels much below the maximum contaminant level goal of 0.8 mg/L (Hynes, and Knowles 1983). A field study of Texas distribution systems employing chlorine dioxide for taste and odor control and chloramination for secondary disinfection showed that even low dosages of ClO_2^- (0.05 mg/L) can inactivate 3 to 4 logs of AOB over several hours (McGuire, Lieu, and Pearthree 1999).

In addition to these chemical strategies, cleaning the distribution system with a flushing program and reducing water age can be effective long-term measures. Nitrifying bacteria are very slow growing and typically found in reservoirs with long detention times (Ike, Wolfe, and Means 1988; Wolfe et al. 1988; Wolfe et al. 1990) or at the ends of distribution systems (Cunliffe 1991). Scouring deposits to limit biofilm development and avoiding stagnant water can create conditions that are restrictive for the long doubling time of these microbes.

Monitoring Surrogate Parameters

In lieu of directly measuring AOB concentration, a number of researchers have attempted to identify a surrogate parameter that would serve as an indicator of nitrification potential. Parameters that have been studied include temperature, nitrite, nitrate, total chlorine residual, free ammonia, pH, HPCs, and dissolved oxygen (Wolfe et al. 1990; Cunliffe 1991; Lieu, Wolfe, and Means 1993; Wilczak et al. 1996; Smith 2001). While some of these individual studies reported an observed correlation between one or more parameters and nitrification episodes, collectively these efforts have failed to demonstrate a consistent correlation between these potential surrogate indicators and nitrification episodes for predictive purposes. Cunliffe (1991) found that increases in oxidized nitrogen species and decreases in total chlorine residuals correlated with increased numbers of AOB. However, Wolfe *et al.* (1990) generally observed poor correlations between AOB numbers and nitrite and chlorine concentrations, and an increase in nitrite concentration has been reported to lag behind AOB growth (Ike, Wolfe, and Means 1988; Lieu, Wolfe, and Means 1993). Wolfe *et al.* (1990) reported a strong correlation with temperature and with HPCs at one location. Cunliffe (1991) evaluated and did not observe a correlation with temperature. A broad survey of chloraminating utilities (Wilczak et al. 1996) reported the lack of consistent and reliable correlation between nitrification and specific indicator parameters. This report noted that changes in these parameters do not provide conclusive evidence of nitrification, as they can also be associated with other processes.

AMMONIA-OXIDIZING BACTERIA

Overview

Nitrification is an important microbial process in aquatic environments. It is a two-step biochemical process involving a sequential conversion of ammonia to nitrite and then nitrate (Equations 2.12 and 2.13 respectively). If the nitrification process is carried out completely through both steps, the elimination of ammonia produces more biologically stable water and reduces the costs associated with additional disinfectant requirements (Crooks et al. 1986). In contrast to the benefits of complete nitrification, incomplete or partial nitrification in the distribution system can adversely affect water quality. Unlike in wastewater treatment, nitrite is often produced in detectable levels due to partial nitrification in chloraminated drinking water systems. Although NOB may play an important role in the ecology of nitrification episodes by affecting the speciation of nitrogen and providing an alternate fate for nitrite than reaction with chloramines, the growth of NOB would not be a suitable monitoring parameter to detect or predict nitrification events. While NOB populations have been detected in chloraminated systems (Regan, Harrington, and Noguera 2001; Regan, Harrington, and Noguera 2002), there is evidence from laboratory reactors that their growth may have a substantial lag behind that of AOB (Regan 2001). Therefore, the focus of this study was exclusively on AOB.

AOB are gram negative, chemolithoautotrophic aerobes. They use oxygen as the terminal electron acceptor and ammonia as the electron donor in their energy-yielding reaction, and obtain carbon through the reduction of carbon dioxide via the Calvin cycle (Wood 1986). Their conversion of ammonia to nitrite is a two-step reaction. The first enzymatic step, the oxidation of ammonia to hydroxylamine (NH_2OH), is catalyzed by the enzyme ammonia monooxygenase (McTavish, Fuchs, and Hooper 1993), and then hydroxylamine is oxidized to nitrite by hydroxylamine oxidoreductase (Sayavedra-Soto, Hommes, and Arp 1994).

Initial AOB taxonomy was defined by morphologic distinction, leading to the description of five genera: *Nitrosomonas*, *Nitrosococcus*, *Nitrosospira*, *Nitrosovibrio*, and *Nitrosolobus* (Bock et al. 1991). Over the past decade, the description of AOB diversity has been revised based on comparative gene sequence analyses. AOB can be classified as two monophyletic groups based on a comparison of 16S rDNA sequences (Teske et al. 1994). One group is within the γ subdivision of the *Proteobacteria* and includes *Nitrosococcus oceani* and *Nitrosococcus halophilus*. The other group, which includes all known freshwater AOB, belongs to the β subdivision of the *Proteobacteria* and subdivides into five different clusters: *Nitrosomonas marina*, *Nitrosomonas oligotropha*, *Nitrosomonas europaea-Nitrosococcus mobilis*, *Nitrosomonas communis*, and *Nitrosospira* (Purkhold et al. 2000). The genera *Nitrosospira*, *Nitrosolobus* and *Nitrosovibrio* have been reclassified into the single genus *Nitrosospira* (Head et al. 1993; Teske et al. 1994; Stehr et al. 1995). The 16S rDNA sequence relationships are graphically depicted in Figure 2.3. The alpha subunit of the functional gene ammonia monooxygenase (*amoA*) has also been used to characterize AOB diversity (Rotthauwe, Witzel, and Liesack 1997; Horz et al. 2000; Park, Regan, and Noguera 2002). Based on comparative *amoA* sequences of the β-subclass AOB, Purkhold *et al.* (2000) suggested three clusters, the *Nitrosospira* cluster, the *Nitrosomonas marina* cluster that includes *Nitrosomonas oligotropha* and *Nitrosomonas ureae*, and the *Nitrosomonas europaea-Nitrosococcus mobilis* cluster.

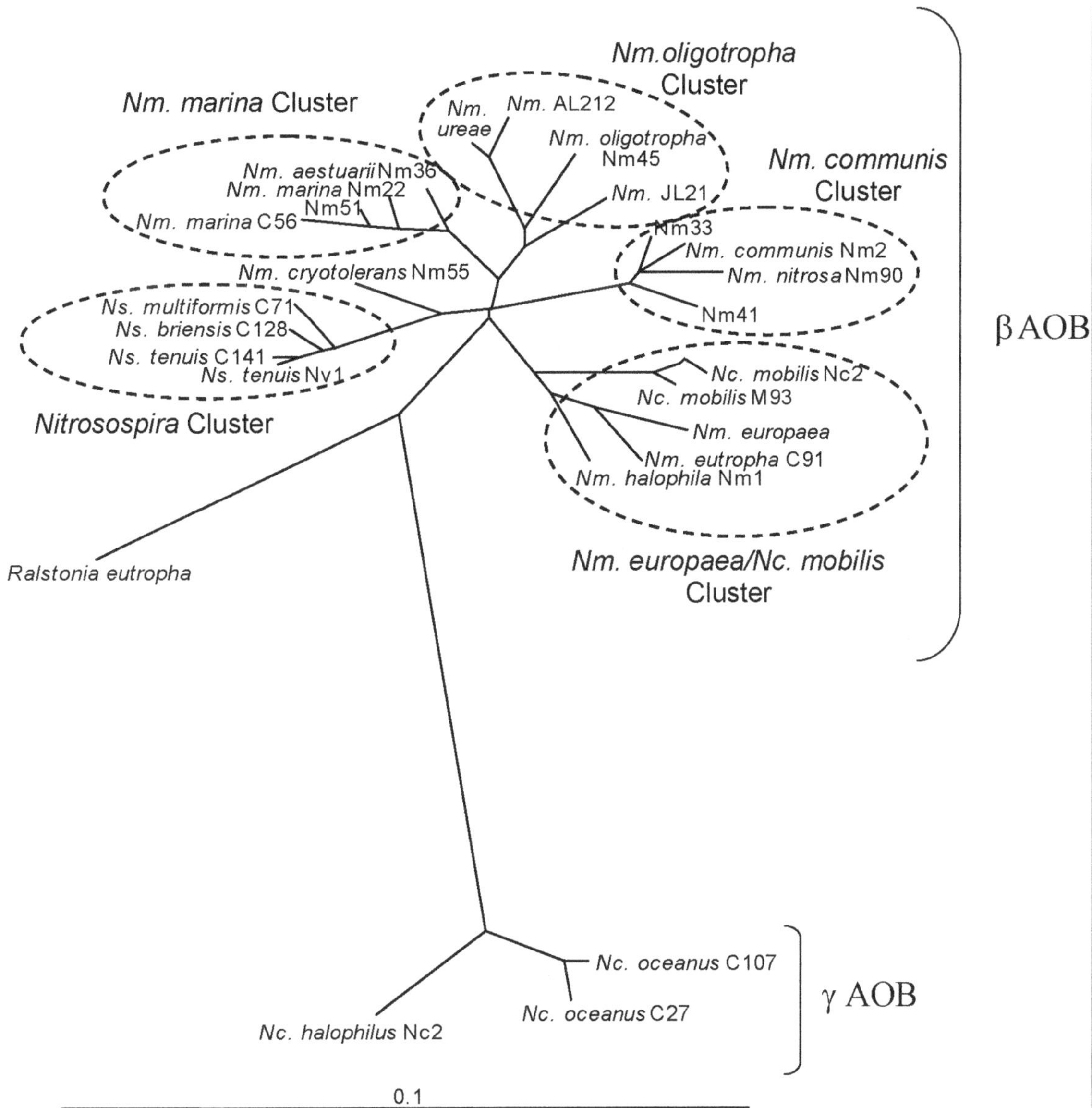

Figure 2.3 Comparative analysis of AOB 16S rDNA sequences

Culture-Based Measurements

The water industry has traditionally employed the most-probable number (MPN) technique for determining AOB concentrations (Wolfe et al. 1990; Cunliffe 1991; Lieu, Wolfe, and Means 1993). This is a culture-based method that involves replicated serial dilutions of a sample in an ammonia mineral medium, incubation under defined conditions of temperature and duration, and testing each dilution series for the formation of nitrite associated with AOB growth. The AOB concentration is then estimated using the number of nitrite-positive tubes at each dilution and a MPN table appropriate for the number of replicates and extent of dilution (APHA, AWWA, and WEF 1995). It is common to use a five-tube procedure with Soriano and

Walker medium (Soriano, and Walker 1968), with incubation in the dark at 28 to 30°C for three to four weeks.

There are several problems, both operational and technical, associated with quantifying AOB using the MPN method. AOB are very slow growing bacteria due to the poor energetics associated with using ammonia as an energy source and carbon dioxide as a carbon source. This necessitates using extended incubation periods to allow the detection of AOB growth. Although three (Wolfe et al. 1990) to four (Cunliffe 1991) weeks are typically used, it has been shown that MPN concentration estimates for AOB can continue to increase over an incubation period of 55 days (Matulewich, Strom, and Finstein 1975). MPN estimates have also been shown to underestimate AOB numbers, accounting for as little as 0.1 to 5.0% of the actual population (Belser, and Mays 1982). This is perhaps due in part to the selective nature of enrichments and the inability to recover AOB that do not grow in the mineral medium (Juretschko et al. 1998). Another possible explanation for these poor recovery efficiencies is the tendency of AOB to form cell aggregates (Watson et al. 1989; Wagner et al. 1995), which precludes accurate cell enumeration with a dilution-to-extinction method such as the MPN technique. Hence, MPN-based enumeration of AOB does not provide data in an operationally relevant timeframe, and the results are subject to underestimating the true AOB numbers.

Molecular Protocols

Recent advances in the development of rapid, culture-independent molecular detection tools and the advent of quantitative capabilities in these tools provide a new opportunity to explore the direct use of AOB numbers as a nitrification indicator and monitoring parameter in place of strictly surrogate parameters. There has been considerable interest in applying molecular techniques to the study of nitrifying bacteria due to their prohibitively slow growth rates and the failure of culturing to represent the true diversity of nitrifier communities. The majority of this research has focused on the development and application of nucleic acid-based techniques for the specific detection and characterization of nitrifiers in soil, sediment, wastewater, and water environments. After only a decade of application, these molecular-based studies collectively have changed many of the paradigms of nitrifying communities that had become established based on culture-based research efforts over the preceding century. Included in these molecular studies is the recent characterization of nitrifying communities in chloraminated distribution systems (Regan, Harrington, and Noguera 2001; Regan, Harrington, and Noguera 2002). These studies found the exclusive presence of AOB related to *Nitrosomonas oligotropha* and the ubiquitous detection of NOB from the genus *Nitrospira*, two nitrifier groups that until recently had been largely overlooked in freshwater environments. Nucleic acid detection and identification techniques can be broadly categorized as variations of either fluorescent in situ hybridization (FISH) or polymerase chain reaction (PCR) approaches. The following sections elaborate on these techniques, with an emphasis on protocols for rapid enumeration.

Fluorescent In Situ Hybridization

FISH is a whole-cell technique that involves the specific hybridization of a fluorescently labeled oligonucleotide probe to the ribosomal RNA (rRNA) of target organisms (Amann, Krumholz, and Stahl 1990). rRNA is an abundant molecule in active cells, allowing sufficient probe-conferred fluorescence within target cells to allow their discrimination from non-target organisms. FISH has been extensively used to study AOB communities in a variety of environments, including laboratory-scale reactors and environmental samples (Wagner et al. 1995; Mobarry et al. 1996; Pommerening-Röser, Rath, and Koops 1996; Schramm et al. 1996; Wagner et al. 1996; Juretschko et al. 1998; Schramm et al. 1998; Schramm et al. 1999; Okabe, and Watanabe 2000; Gieseke et al. 2001). It can be used to provide information on cell identity, abundance, and spatial relationships in environmental samples. Figure 2.4 illustrates AOB-targeted FISH with a pure culture and an intact activated sludge floc.

FISH is an inherently quantitative method; however, the microscopy-based enumeration of target organisms can be a prohibitively time-consuming effort. Flow cytometry (FCM) offers an alternative and rapid approach for enumerating FISH preparations in a high-throughput manner (Amann et al. 1990; Wallner, Erhart, and Amann 1995; Porter, and Pickup 2000). Unlike microscopy-based analysis, FCM can count thousands of target cells and non-target particles in seconds. As illustrated in Figure 2.5, the sample is injected into a hydrodynamic flow cell through which a focused laser beam is transmitted. Sensors detect light scattering and fluorescence emission associated with particles in the sample. In principle, target cells exhibit an increase in fluorescence intensity above background levels, enabling their discrimination from other particles. The target cell counting rate can be converted to a cell concentration using an internal standard of fluorescent beads of known concentration. FCM has been used to analyze drinking water samples for the presence of specific microorganisms using immunofluorescence, primarily targeting *Cryptosporidium* (Vesey et al. 1997) but also looking at different AOB serotypes (Völsch et al. 1990). However, there are no published reports of combining AOB-targeted FISH with FCM.

A potential problem associated with FISH is a low target-cell signal that precludes a definitive discrimination between target cells and non-target cells or autofluorescent debris. This can be affected by probe design, as rRNA exhibits variable accessibility along its length (Fuchs et al. 1998). Assuming the probe is designed to complement an accessible region of the rRNA, low cell fluorescent intensity can arise from a low rRNA content in inactive cells (Ruimy et al. 1994). This might be expected of AOB growing in drinking water distribution systems, which

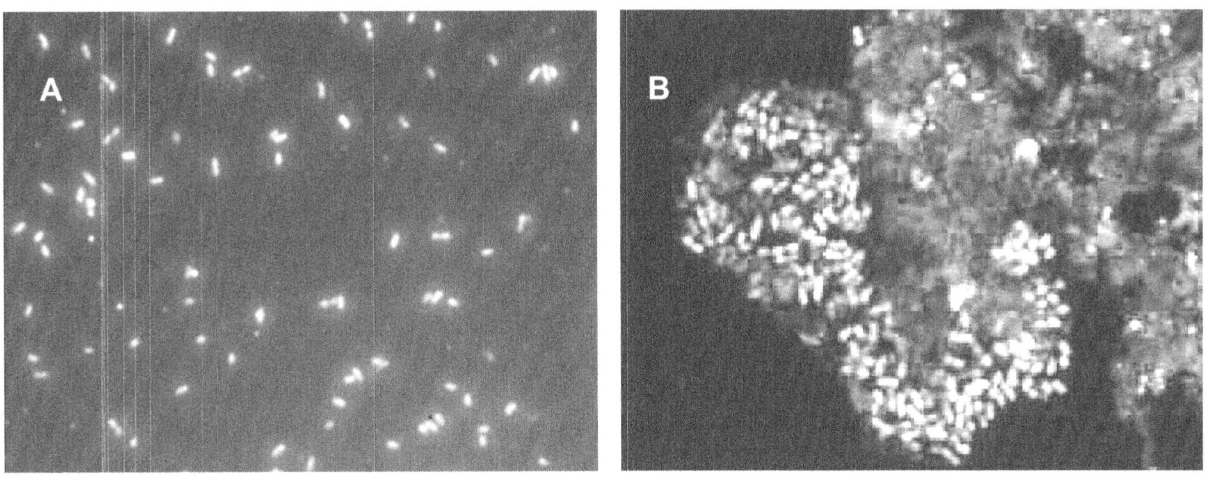

Figure 2.4 Images of ammonia-oxidizing bacteria-targeted fluorescent in situ hybridization: (A) a pure culture of ammonia oxidizers, (B) an intact floc from a nitrifying activated sludge system

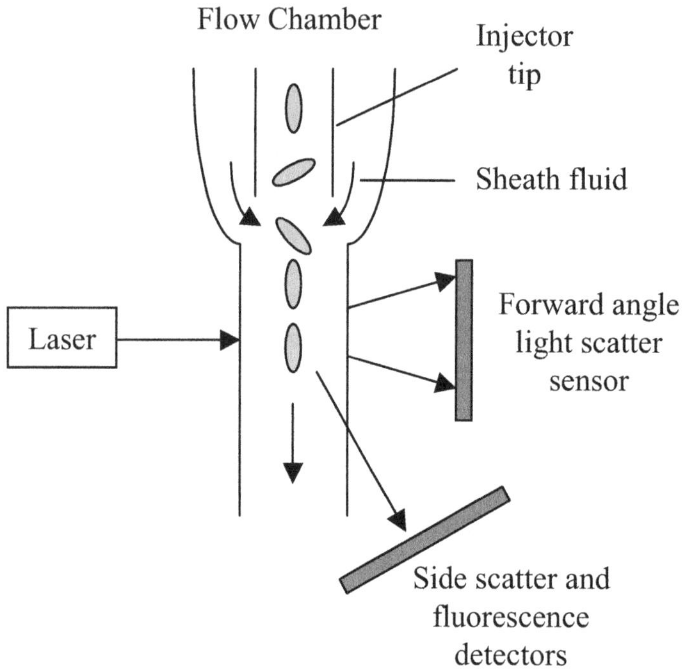

Figure 2.5 Schematic of the flow chamber and particle detection components of a flow cytometer

would be growing in substrate-limited conditions. Strategies to overcome this limitation include the hybridization of multiple probes or probes with multiple fluorochromes, or the use of an enzymatic amplification strategy.

Tyramide signal amplification (TSA), or catalyzed reporter deposition (CARD), is one such enzymatic approach for increasing the fluorescence of target cells (Not et al. 2002; Pernthaler, Pernthaler, and Amann 2002; Biegala et al. 2003; Sekar et al. 2003). This method uses oligonucleotide probes labeled with the enzyme horseradish peroxidase (HRP) instead of a fluorochrome. After hybridizing the HRP-labeled probe to the target rRNA and washing unbound probe, fluorochrome-conjugated tyramide is added to the sample. HRP catalyzes the fixation of radicalized tyramide, leading to the deposition of multiple tyramide-associated fluorochromes in the vicinity of each probe. Relative to traditional FISH, this variation requires some additional time associated with cell permeabilization to allow HRP penetration into the cell and incubation to allow HRP activation of the tyramide. The benefit is enhanced fluorescence intensity, reportedly 10 to 20 times higher than single-fluorochrome probes (Not et al. 2002).

Polymerase Chain Reaction

PCR is an enzymatic method of dramatically amplifying the concentration of a target DNA fragment, thereby facilitating subsequent detection and analysis. Most sections of DNA are present in a single copy or several copies on the chromosome, making direct detection infeasible. PCR amplifies the target sequence using the *Taq* DNA polymerase enzyme and a pair of oligonucleotide primers that specifically complement sites flanking the DNA region of interest (Figure 2.6). It is a cyclic reaction consisting of three basic temperature-controlled steps:

denaturation, in which the template DNA is heated to between 94 and 97°C to separate the two strands of the target DNA; annealing, in which the reaction temperature is lowered to allow the primers to specifically bind with their respective targets on the separated single strands of the template DNA; and elongation, during which the temperature is raised to 72°C, allowing *Taq* DNA polymerase to extend from the primers and generate a complementary copy of the template DNA. Each successive cycle theoretically doubles the template DNA concentration, producing a geometric increase in fragment copy number with cycle number. Detection of the amplification products can be performed by DNA staining or the use of a fluorescently labeled primer. There are also a number of protocols, many of which have been applied to AOB, for processing PCR amplification products prior to detection to generate additional community information.

Although there is a theoretical geometric increase in PCR product concentration with cycle number (Figure 2.6), which would allow the estimation of the initial DNA template concentration, traditional PCR-based protocols are at best semi-quantitative due to methodological issues that can introduce quantitation errors in PCR-generated data (Wintzingerode, Göbel, and Stackebrandt 1997). These can include a decrease in polymerase efficiency over the course of multiple cycles, the depletion of primers and deoxynucleoside triphosphate precursors, differential amplification efficiency of various templates, and inhibition of amplification efficiency by matrix components present in the sample. Within the past several

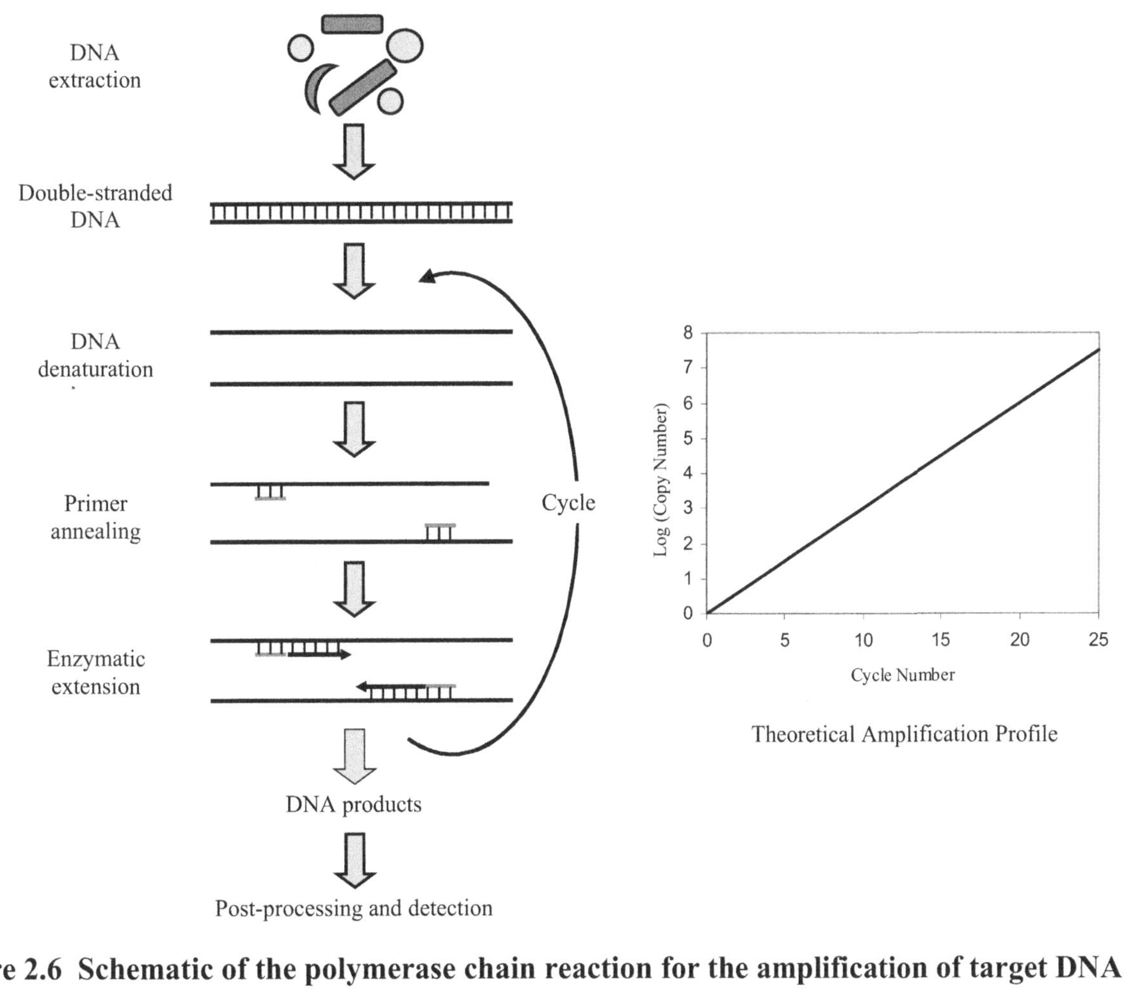

Figure 2.6 Schematic of the polymerase chain reaction for the amplification of target DNA

years, PCR techniques have been developed that provide quantitative data, including limiting-dilution PCR (Sykes et al. 1992), kinetic PCR (Higuchi et al. 1993), competitive PCR (Felske, Akkermans, and DeVos 1998; Mendum, Sockett, and Hirsch 1999), and real-time PCR.

The principle of real-time PCR is to quantify the exponential formation of amplification products and back-calculate the initial template concentration. Amplification products are quantified by monitoring changes in the intensity of a fluorescent signal that is proportional to product concentration. This proportionality can be achieved in several ways, the simplest of which involves the use of a nonspecific fluorescent dye that intercalates in the DNA double strand and gives an increased fluorescent signal when bound to the double-stranded DNA amplification products. The main disadvantage of this method is its susceptibility to false positives due to nonspecific PCR products or primer-dimer formation, since the dye has no DNA sequence specificity. An alternative approach is to use a TaqMan probe specific for a region of the amplification product between the two primers. The probe is labeled on one end with a fluorochrome and on the other end with a molecule that quenches the fluorescent emission when it is in proximity to the fluorochrome. During amplification of target DNA, the 5'-exonuclease activity of DNA polymerase cleaves the probe, giving rise to an increase in fluorescence as the quencher is no longer neighboring the fluorochrome. This latter approach is illustrated in Figure 2.7. Recently a real-time PCR protocol was developed specifically for AOB enumeration (Hermansson, and Lindgren 2001). However, this protocol requires some specific modifications to make it suitable for measuring AOB in chloraminated drinking water systems.

The advantage of real-time PCR is that it focuses on the exponential phase of PCR product accumulation rather than on the end products. Therefore, it is less affected by changes in amplification efficiency that occur over sequential PCR cycles. In addition, real-time PCR can measure the amount of target DNA over a dynamic range of around six orders of magnitude. Quantification is achieved by measuring the increase in the fluorescent signal during the exponential phase of PCR and determining the cycle threshold (Ct) value, which represent the cycle at which there is an increase in target fluorescence above a threshold fluorescent intensity. The Ct value is inversely proportional to the copy number of the target gene or cell numbers. With the use of a standard curve correlating Ct values to known target cell or DNA concentrations, the target DNA concentration in an unknown can be estimated.

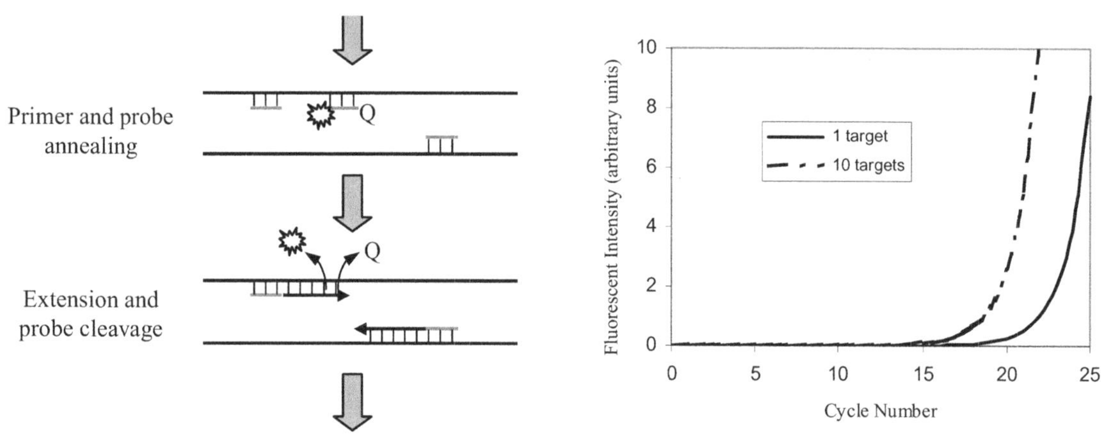

Figure 2.7 Abbreviated diagram of real-time polymerase chain reaction using a TaqMan probe showing the cleavage of the probe and release of the fluorochrome and quencher molecule

CHAPTER 3
MATERIALS AND METHODS

PURE CULTURES

Cultures of *Nitrosomonas europaea* (ATCC 19718) and *Nitrosospira multiformis* (ATCC 25196) were obtained from American Type Culture Collection (ATCC), Manassas, Virginia, for use as positive control AOB cells. They were grown in the dark at 28 ºC on a shaker using ATCC culture media 2265 and 438, respectively (Table 3.1). Cultures were checked daily for pH by means of a color indicator, and adjustments were made as necessary with sterile 5% K_2CO_3 to maintain the pH between 7.2 and 7.8. These cultures were transferred to new media every six to eight weeks.

Table 3.1
ATCC growth media for AOB

#2265 for *Nitrosomonas europaea*	#438 for *Nitrosospira multiformis*
Solution 1:	Solution 1:
$(NH_4)_2SO_4$, 4.95 g KH_2PO_4, 0.62 g $MgSO_4$ (1.0 M stock solution), 1.1 mL $CaCl_2$ (1.0 M stock solution), 0.3 mL $FeSO_4$ (30 mM in 50 mM EDTA), 0.5 mL $CuSO_4$ (50 mM stock solution), 0.02 mL Distilled water, 1.2 L *Filter sterilize*	$(NH_4)_2SO_4$, 1.65 g KH_2PO_4, 0.087 g $MgSO_4 \cdot 7H2O$, 0.2 g $CaCl_2 \cdot 2H2O$, 0.02 g Versene, 1.0 mg
Solution 2:	Solution 2 (trace element):
KH_2PO_4, 8.2 g NaH_2PO_4, 0.7 g Distilled water, 300 mL *Bring to pH 8.0 with 10 N NaOH, filter sterilize*	$Na_2MoO_4 \cdot 2H_2O$, 0.1 mg $MnCl_2 \cdot 4H_2O$, 0.2 mg $CoCl_2 \cdot 6H_2O$, 2.0 mg $CuSO_4 \cdot 5H_2O$, 0.02 mg $ZnSO_4 \cdot 7H_2O$, 0.1 mg
Solution 3:	Solution 3:
Na_2CO_3 anhydrous, 0.6 g Distilled water, 12 mL *Filter sterilize*	Phenol red, 5.0 mg 3-5% K_2CO_3
	Total solution volume: 1 L *Autoclave and adjust pH to 7.5*
Combine after sterilizing separately	

Source: Adapted from American Type Culture Collection web site (www.atcc.org)

Pure culture negative controls of *Ralstonia basilensis* (DSM 11853) and *Escherichia coli* (ATCC 11775) were used for contrast with target AOB cells. *Ralstonia,* a β–Proteobacterium like the AOB strains, was selected because of its rRNA sequence similarity to AOB. *E. coli* was selected as a representative of a dissimilar sequence. *R. basilensis* was grown in nutrient broth and *E. coli* cultures were grown in LB. Both cultures were grown overnight in their respective media and were incubated in a shaker at 28 °C.

FULL-SCALE SAMPLING

During the first season of full-scale sampling, samples were collected routinely from four chloraminated systems: San Francisco Public Utilities Commission (SFPUC) from February 17 to September 22, 2004; Ann Arbor Water Utilities (AAWU) from June 29 to November 1, 2004; Pinellas County Utilities (PCU) from July 22 to November 2, 2004; and the Alameda County Water District (ACWD) from June 9 through December 21, 2004. In addition, a single sampling event was conducted at Louisville Water Company (LWC) in August 2004. The SFPUC system was sampled at two reservoir locations, Sunset North and Stanford Heights, beginning just two weeks after the system conversion to chloramines on February 2, 2004. Sunset South was added to the sampling locations in June 2004. Samples were collected from four tap locations in the AAWU system. In the PCU system, which switched to chloramines in May 2002, four distribution system locations were included in the sampling. Maintenance of this system during the sample collection period included distribution system flushing, breakpoint chlorination that commenced on October 25, 2004, and draining and modification of the Gulf Beach tank during August and September 2004 to reduce the water age in this above-ground storage tank. In the ACWD system, samples were collected from their 15 MG finished water storage reservoir located in Fremont, California. Samples were collected at reservoir ports 2, 4, and 5, as designated in Figure 3.1. ACWD was not initially included as a participating utility in this study, but it was added when the other systems did not show any chemical or biological evidence of nitrification. The Alameda Reservoir has historically experienced a reduction in disinfectant residual as a result of nitrification, even during the cold temperature season. Our monitoring of this reservoir spanned several nitrification episodes, along with ACWD's response of breakpoint chlorination on June 15, July 12, July 27, July 29, and August 31, 2004. The District also explored the implementation of a ClorTec Reservoir Management System™ (RMS), which adds and mixes chlorine and ammonia to boost the chloramine residual. This booster chloramination system began on September 27, 2004, was turned off on October 16, 2004 and resumed on November 11, 2004.

Additional sampling was performed in the second season at SFPUC and the Washington DC Water and Sewer Authority (DCWASA). In both utilities, a closed-loop section of the distribution system that was known to regularly experience nitrification was sampled. SFPUC provided samples from their Metershop tank, which is a closed-loop recirculating system (~17,500 gallons) on a corporation yard that is used to calibrate water meters, from July 21 to August 12, 2005. The system was sampled in two locations, in the tank and at the tank inlet (Figure 3.2). DCWASA was added to the full-scale sampling to monitor a school that was experiencing nitrification during the summer months. Samples were collected from the distribution main at a hydrant outside the school building (Figure 3.3) and from the school kitchen sink (Figure 3.4) between June 6 to August 15, 2005. The school system was periodically flushed by leaving the kitchen sink tap open.

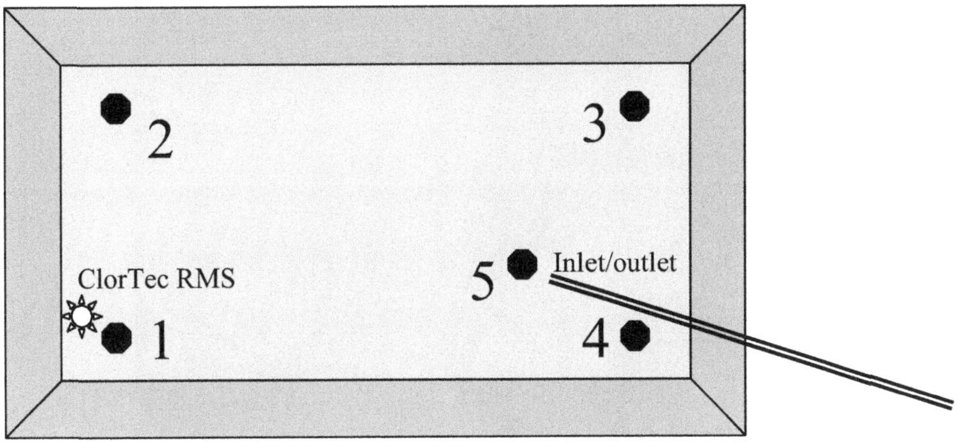

Figure 3.1 Locations of sampling ports in Alameda County Water District Alameda Reservoir

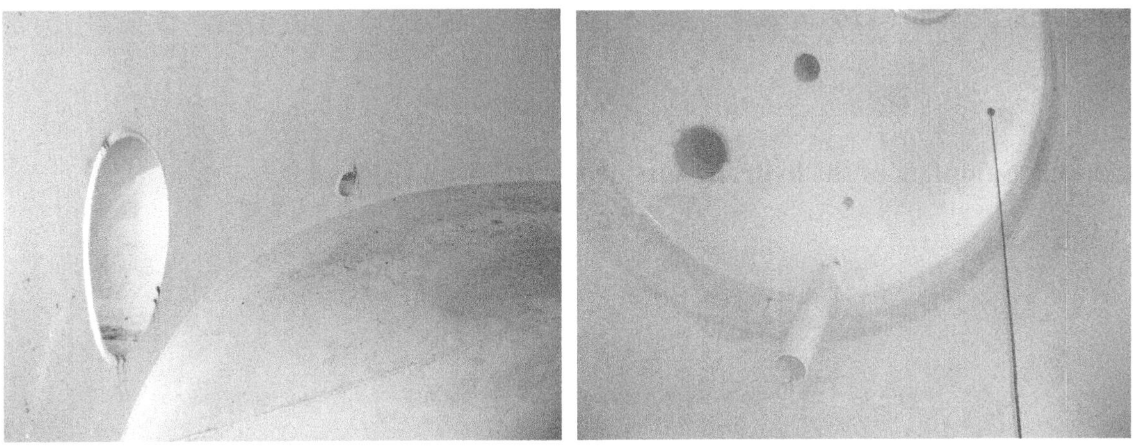

Figure 3.2 The Metershop tank recirculation in the SFPUC system (showing outlet (left) and inlet (right) when tank is empty)

Figure 3.3 Sampling of the hydrant outside the DCWASA school

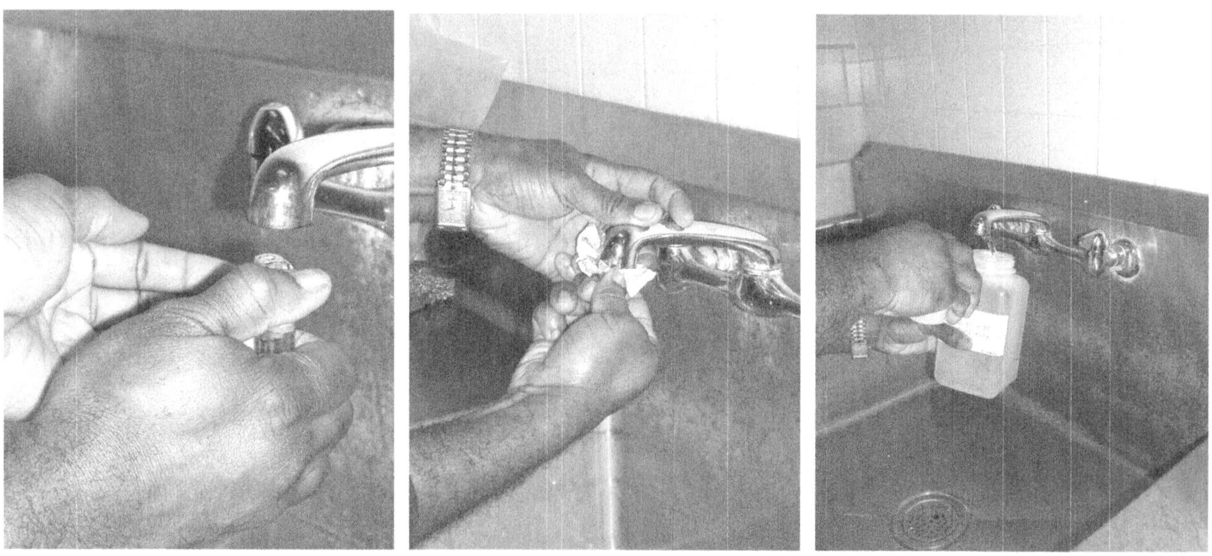

Figure 3.4 Sampling of the kitchen sink in the DCWASA school. The tap was flamed and swabbed with alcohol prior to sampling

In each of the systems, duplicate samples were collected in sterile 1-L bottles containing sodium thiosulfate and were shipped on ice to the Pennsylvania State University. Samples were analyzed upon arrival for AOB using the traditional most-probable number (MPN) assay and the real-time polymerase chain reaction (PCR) protocol described below. In addition, samples were analyzed by each system for a suite of chemical parameters (varied by utility). Total chlorine, nitrite, nitrate (at PCU), monochloramine, total ammonia, free ammonia, pH, and temperature were analyzed at the time of sample collection. Total organic carbon (TOC), nitrate (at AAWU), and heterotrophic plate count using R2A agar (HPC-R2A) were analyzed in the respective laboratory. Table 3.2 lists the methods for all parameters included in the sampling plan. Exceptions to this include the single sampling event from LWC, in which nitrite was estimated using Hach test strips.

FLUORESCENT IN SITU HYBRIDIZATION (FISH)

Probe Design

FISH probes were designed to specifically target AOB using 16S rRNA sequences retrieved from the National Center for Biotechnology Information (NCBI) and aligned using ClustalX (Thompson et al. 1997). The analysis was performed on 42 *Nitrosomonas* sequences and 39 *Nitrosospira* sequences. The alignment was investigated for sequence regions for which all AOB exclusively have the same sequence. In addition, *E. coli* and *R. basilensis* sequences were aligned with AOB as negative control cells. The theoretical specificities of potential probe sequences with matches to aligned AOB and mismatches to the non-AOB sequences were

Table 3.2
Water quality analyses

Parameter	Method
Total Chlorine	Hach – 8167 (Standard Methods 4500-Cl G)
Monochloramine	Hach – 10200 (Indophenol method)
Total Ammonia-N	Hach – 8155 (Salicylate method)
Free Ammonia-N	Hach - 10200 (Indophenol method)
Nitrite	Hach - 8507 (Diazotization method)
Nitrate	AAWU: Standard Methods 4500-NO_3^- E PCU: Hach – 8192 (Cd reduction method)
HPC-R2A	PCU, ACWD: Standard Methods 9215 B (7 day) AAWU: Standard Methods 9215 C (7 day)
TOC	Standard Methods 5310
pH	Standard Methods 4500-H^+ B
Temperature	Standard Methods 2550 B

further evaluated using NCBI's Basic Local Alignment Search Tool (BLAST) (Altschul et al. 1990).

Based on these sequence analyses, probes Nso1225 (Mobarry et al. 1996; Mobarry et al. 1997) and AOB770 were selected for experimental evaluation. Nso1225 is commonly used for detecting AOB and is an exact complement to all known β–*Proteobacterial* AOB except *Nitrosococcus mobilis*, with which there is one nucleotide mismatch near the 5' end. AOB770 was designed in this study, and the FISH hybridization and washing conditions were optimized with regard to formamide and corresponding salt concentrations, respectively. Note that these probes do not target AOB within the γ-subclass of *Proteobacteria*, which presently includes only marine species of *Nitrosococcus*. EUB338 was used as positive control probe, NonEUB338 was used to characterize non-specific binding, and a no-probe control was performed to characterize autofluorescence. Probe sequences and FISH conditions are listed in Table 3.3. All probes were manufactured by Sigma-Genosys. For the initial FISH work involving microscopic analysis, probes were 5' labeled with sulfo-indocyanine Cy3. However, this was later changed to FITC to allow compatability with a flow cytometer equipped with a single argon-ion laser emitting light at 488 nm.

Protocol for Microscopy Analysis

The FISH protocol used in this study for subsequent microscopic visualization (presented in the Appendix) was adapted from Mobarry et al (1996) and Ootsubo et al (2002). Because of their slow growth rates, pure cultures of AOB were collected for FISH a few weeks after their transfer to new media. AOB cultures (5 μL) were spotted on each well of multi-well slides with heavy Teflon coating (Erie Scientific, Portsmouth, NH), air dried, and fixed in 3 volumes (i.e., 15 μL) of 4% paraformaldehyde for 30 min. The slides were washed with 2 mL of PBS and air dried. Non-AOB cells were grown overnight in their respective liquid medium, and aliquots were transferred to microcentrifuge tubes and fixed with paraformaldehyde for 30 min. These cells were then pelleted using an Eppendorf microcentrifuge at 13,200 rpm for 2 min. (AOB cell cultures were spotted directly on slides due to poor recovery using centrifugation.) The pellets of non-AOB cells were then washed and resuspended in one volume of PBS, and bacterial suspensions (3-μL aliquot of fixed cells) were spotted onto slide wells. After air drying, cells were dehydrated by serial immersion of slides in 50%, 80%, and 96% ethanol for 2 min each.

Table 3.3
FISH oligonucleotide probes used in this study

Probe	Sequence (5'-3')	16S rRNA Target Specificity	% F*	Ref.
Nso1225	CGC CAT TGT ATT ACG TGT GA	AOB (β–subclass)	35	(Mobarry et al. 1997)
AOB770	CTG AGC GTC AGT GTY AAC CC	AOB (β–subclass)	35	This study
EUB338	GCT GCC TCC CGT AGG AGT	Domain Bacteria	20	(Schramm et al. 1996)
NonEUB338	ACT CCT ACG GGA GGC AGC	Non sense	20	(Wallner, Erhart, and Amann 1995)

* Concentration of formamide in the hybridization buffer

Slide-adhered fixed cells were then subjected to probe hybridization and washing. A 16 μL aliquot of hybridization buffer containing 35% formamide (Nso1225 and AOB770) or 20% formamide (EUB338 and nonEUB338), 0.9 M NaCl, 20 mM Tris-HCl (pH 7.2), 0.01% sodium dodecyl sulfate (SDS), and 50 ng of labeled oligonucleotide probe was spotted on each well. Slides were incubated for 90 min at 46°C in moisture chambers equilibrated to 0.9 M NaCl. In addition to the single-probe hybridizations, experiments were also conducted using dual labeling with Nso1225 and AOB770 to increase fluorescence of target AOB cells. These probes are compatible in terms of hybridization and washing conditions, allowing them to be added simultaneously. After hybridization, the slides were washed with 2 mL of pre-warmed wash buffer, immersed for 20 minutes in 40 mL of wash buffer at 48°C, rinsed with 2 mL of distilled water, and air dried. Wash buffer consisted of 47 mM NaCl (Nso1225 and AOB770 probes) or 170 mM NaCl (EUB338 and nonEUB338 probes), 20 mM Tris-HCl (pH 8.0), and 0.01% SDS. Cells were counterstained using 15 μL of 4',6-diamidino-2-phenylindole (DAPI, 0.5 ng/μL), an intercalating DNA stain, for 5 minutes. The excess DAPI was washed off using 2 mL of distilled water. After air drying, slides were mounted with 1,4-diazobicyclo[2.2.2]octane (DABCO) antifade solution containing 25 mg DABCO/mL of PBS:glycerol (1:9 vol/vol), and coverslips were sealed to the slides using clear nail polish. The probe-conferred signals were visualized using a Zeiss Axiophot epifluorescent microscope with appropriate filter sets and a Zeiss AxioCam CCD (charge-coupled device) black and white camera. Average fluorescent intensities were calculated by measuring cell intensities under the various experimental conditions from images obtained using identical exposure times.

Protocol for Flow Cytometric (FCM) Analysis

The FCM-FISH protocol in this study (presented in the Appendix) was adapted from Amann et al. (1990), Jaroszeski and Heller (1998), and Baerlocher et al (2002). The principles are the same as microscopy-based FISH, except that microscopic analysis requires cells adhered to a slide while FCM analysis is performed on cell suspensions. Initial efforts entailed sequential centrifugation and resuspension of cells in each successive component of the FISH protocol (i.e., fixative, hybridization buffer, and wash buffer). However, given the poor pelleting achieved with the AOB pure cultures and the expectation of very dilute suspensions from distribution system samples, there was a concern about accumulating significant loses in AOB numbers with this approach. To address this, the protocol was modified by replacing centrifugation with filtration to allow the transfer of cells to successive components. This FCM-FISH protocol section is divided into two methods, one for non-target cells and one for target AOB cells.

The non-target cells *E. coli* and *R. basilensis* were fixed as described above, and the PBS-resuspended cells were mixed with ethanol (1:1 vol) and stored at -20°C until further processing. Hybridizations were performed in microcentrifuge tubes containing 50 μL of prepared sample, 400 μL hybridization buffer with the same composition as provided above for microscopy-FISH, and 20 μL of 50 ng/μL probe solution. The suspended hybridization step was performed for 3 hrs at 46°C. After hybridization, suspended cells were pelleted by centrifugation at 13,200 rpm for 3 minutes and resuspended in 800 μL wash buffer for 20 min at 48°C. This wash process was repeated once. Then the pellet was resuspended in at least 500 μL cold wash buffer and kept on ice prior to FCM analysis.

A filtration-based separation and transfer method was used for the AOB pure cultures, which had very low cell densities. A 100 to 1000 μL sample of pure culture cells was filtered using a 0.22 μm polycarbonate filter (Poretics, Osmonics Inc.). The filter paper was then carefully removed and transferred to a 2 mL microcentrifuge tube with 1 mL of TE buffer (10mM Tris, 1.0 mM Na_2EDTA, pH 7.5). The cells were detached from the filter into the buffer solution through vortexing for 1.5 to 2 minutes. After vortexing, the filter was checked to confirm that cell detachment was effective using DAPI staining and epifluorescent microscopy. The dispersed cells were fixed by combining 500 μL with 1500 μL of 4% paraformaldehyde solution and incubating for 30 minutes. The fixed cells were filtered with a 0.22 μm polycarbonate filter, which was then washed two or three times with 2 mL of PBS. The filter was removed from the filter apparatus, transferred to a microcentrifuge tube with 500 μL of PBS, and vortexed to detach the cells. Resuspended cells were filtered, resuspended in cold absolute ethanol, and stored at -20°C for subsequent preparations. The hybridization step was performed as described above for non-target cultures. Following hybridization, the suspension was filtered, and the filter was washed several times with 2 mL of wash buffer. Finally, the filter was transferred to at least 500 μL cold wash buffer, cells were detached by vortexing, and samples were kept on ice prior to FCM analysis.

Washed cells were transported on ice to Penn State's Center for Quantitative Cell Analysis. The bacterial suspensions (at least 500 μL) were transferred to 10 mL flow cytometer tubes, and Coulter Standard-Brite fluorescent beads were added in the tubes from a stock concentration of 3.74×10^{10} mL^{-1} to allow the calculation of cell concentration. After diluted beads were added in each tube and mixed well, each sample was injected into the flow cytometer. At least 5,000 cells were counted in each evaluation. The suspensions were analyzed using a Coulter XL-MCL equipped with an air-cooled argon ion laser having a single peak at 488 nm. FITC response was measured in the FL1 channel.

The analysis of fluorescence by gating on single cells was done using WinMDI Version 2.8 (Windows Multiple Document Interface Flow Cytometry Application, available at http://flowcyt.salk.edu/software.html). Quantitative fluorescence intensity results from the single-color cell population were analyzed using MFI 3.4 (available at http://www.umass.edu/microbio/mfi/). This program was created to obtain corrected median fluorescence intensities from flow cytometry listmode data files. This software was useful to convert raw data files (LMD; list mode data) into numeral data for statistical analysis. MFI plots any parameter versus event number or time. The analysis in this study was focused on fluorescence and event number. Events were plotted according to their sequence in the file, not according to the time when they passed the laser.

Tyramide Signal Amplification Protocol

Microscopy Analysis

Due to only a modest probe-conferred fluorescence increase of AOB relative to non-target cells (presented and discussed in Chapter 4), even with the simultaneous hybridization of two probes, attempts to increase target cell fluorescence were pursued through the use of tyramide signal amplification (TSA), or catalyzed reporter deposition (CARD) (Pernthaler, Pernthaler, and Amann 2002; Biegala et al. 2003; Sekar et al. 2003). This approach involves the 5'-end labeling of an oligonucleotide probe with horseradish peroxidase (HRP) *in lieu* of a fluorochrome. Following probe hybridization to target rRNA and washing of unbound probe, fluorochrome-conjugated tyramide is added to the sample. HRP in probe-positive cells radicalizes tyramide, which then complexes with intracellular molecules and results in the deposition of multiple fluorochromes in the vicinity of each HRP-labeled probe. TSA-FISH is reportedly capable of increasing cell intensities by 10 to 20 times over single-labeled probes (Not et al. 2002).

The TSA-FISH protocol consisted of fixation with paraformaldehyde, permeabilization with lysozyme to allow entry of the larger HRP-labeled probe, hybridization, and tyramide signal amplification. A FITC-conjugated tyramide kit (TSA™ Fluorescein System, NEL701) was purchased from Perkin Elmer Life Sciences (Boston, MA). HRP-conjugated oligonucleotide probes of Nso1225 and EUB338 (Table 3.3) were synthesized by Biosource International (Camarillo, CA). The method was tested on pure cultures of *Nitrosomonas europaea* and *Nitrosospira multiformis*, with *Escherichia coli* serving as a negative control. Nso1225 was used to specifically target AOB cells, and EUB338 was used as a positive control probe. Negative controls were performed using a no-probe condition. A 5-6 μL amount of sample was put on a slide well, air-dried, and fixed with 3 volumes of 4% paraformaldehyde for 30 min. Slides were washed with 2 mL of PBS and air-dried. Lysozyme solution containing 10 mg lysozyme/mL, 0.05 M EDTA, and 0.1 M Tris-HCl (pH 7.5) was added for 30 minutes at 37°C. After this step, slides were rinsed with distilled water and dehydrated in 96% ethanol for 1 min. The hybridization step was performed for 2-3 hrs at 35°C in moisture chambers. This incubation temperature was lower than that of regular FISH to maintain HRP activity. A mixed solution of 20 μL hybridization buffer and 2 μL probe was added to each well. The hybridization buffer was composed of 50-60% formamide, 0.9 M NaCl, 20 mM Tris-HCl (pH 7.5), 10% dextran sulfate, 1% blocking reagent from the kit, 0.05% Triton X-100, and 0.02% SDS. After hybridization, the slides were washed with 2 mL of wash buffer, immersed for 10 minutes in 40 mL of wash buffer, and then rinsed with 2 mL of sterile water. Wash buffer consisted of 3 mM NaCl, 20 mM Tris-HCl (pH 7.5), 5mM EDTA, and 0.01% SDS.

After the washing step, TNT buffer containing 0.15 M NaCl, 0.1 M Tris-HCl (pH 7.5), and 0.3% Triton X-100 was added to the slides for 15 min at room temperature to equilibrate the probe-delivered HRP. After excess buffer was removed, 20 μL of the substrate mix was put on each well and incubated at 37°C for at least 10 min in the dark. The substrate mix consisted of 1 part FITC-labeled tyramide and 100 parts amplification diluent, both included in the kit. Excess solution was removed and slides were immersed in 40 mL of TNT buffer for 40 min to remove inactivated FITC-tyramide. The slides were washed with distilled water and immersed in 96% ethanol for 1 min. Finally, DAPI counterstaining, slide mounting, and microscopic visualization and analysis were performed as described above.

FCM Analysis

This protocol was simply a combination of the FCM-FISH protocol and the TSA FISH-microscopy protocol.

REAL-TIME POLYMERASE CHAIN REACTION (PCR)

DNA Extraction

DNA was extracted from the control pure cultures and distribution system samples using the Qiagen DNeasy Tissue Kit (Qiagen, Valencia, CA) according to the instructions for gram negative bacteria. After pure AOB cultures were incubated for at least 15 days, 1 mL of each culture medium was filtered through a 0.22 μm polycarbonate membrane filter (Osmonics Inc., Minnetonka, MN). To concentrate field samples, 1-L samples were filtered through a sterile 0.2 μm polycarbonate filter. Filters from pure cultures and field samples were transferred to microcentrifuge tubes containing 700 μL of TE buffer (10mM Tris, 1mM EDTA, pH 8.0) and vortexed for 10 minutes. Cell suspensions were then subjected to four freeze-thaw cycles, and DNA was recovered and purified using the kit as follows. 180 μL of ATL buffer and 20 μL of proteinase K solution were added and incubated for 1 hour at 55°C. Then 200 μL of AL buffer was added and incubated for 10 min at 70°C. After cell lysis, the extracted DNA was mixed with 200 μL of absolute ethanol and transferred to a DNeasy Mini Spin Column. The column was sequentially washed with 500 μL of AW1 buffer and AW2 buffer. Finally, the genomic DNA was eluted with 200 μL of elution buffer AE and stored at -20°C.

Primer and Probe Design

The real-time PCR task involved adapting PCR protocols that have been used for AOB *characterization* to allow the recovery of *quantitative* data. Initially, the 16S rRNA gene (16S rDNA) was selected as the target for PCR because it is present on the chromosome in only a single copy in all characterized AOB (Aakra, Utåker, and Nes 1999), which simplifies enumeration of AOB cells. Real-time PCR targeting AOB 16S rDNA has been previously performed on soil samples (Hermansson, and Lindgren 2001). However, one of the primers used in this previous study has nucleotide mismatches with several AOB, including three mismatches with *Nitrosomonas oligotropha*, the exclusive AOB species that has been documented in chloraminated drinking water systems (Regan, Harrington, and Noguera 2002; Regan et al. 2003). Therefore, the first stage of this real-time PCR task entailed examining aligned sequences to identify PCR primer candidates. Within the additional primer design constraints imposed by real-time PCR, including relatively short amplicon lengths, two primer pairs (AOB620 with AOB770rc, and AOB1150 with Nso1225, Table 3.4) were identified that held potential to specifically amplify AOB 16S rDNA.

Due to problems with the generation of nonspecific PCR products from non-AOB beta-subclass *Proteobacteria* that have similar sequences to the primer pairs, an alternative approach to AOB-targeted real-time PCR was pursued. Instead of using 16S rDNA as the target gene, we investigated using *amoA*, the gene that codes for the catalytic subunit of ammonia monooxygenase. Unlike 16S rDNA, which is present in all bacterial and archael cells, *amoA* is exclusively found in AOB. This reduces the possibility of encountering nonspecific amplification products from environmental samples. Our analysis showed that the primer pair amoA-1F and amoA-2R (Table 3.4) matches the available *amoA* sequences that extend beyond the target sites of these primers (16 *Nitrosomonas*-derived *amoA* sequences and 18 *Nitrosospira*-derived sequences). These primers target a 491-bp fragment corresponding to positions 332 to 349 and 802 and 822 of the open reading frame of *amoA* for *Nitrosomonas europaea*. These primers showed no amplification products when used with a methanotrophic culture, which possesses the particulate methane monooxygenase gene (*pmo*) that has a similar sequence to *amoA* (Rotthauwe, Witzel, and Liesack 1997). Moreover, Mendum et al. (1999) reported that all the environmental clones derived from soil with these *amoA* primers were related to the β-subgroup ammonia oxidizers, and no false positive results were observed.

Table 3.4
Real-time PCR primer and probe sequences

Primer Pair or Probe	Target	Sequence (5'-3')	Reference
AOB620/ AOB770rc	β-AOB 16S rDNA	GTA AGT CAG ATG TGA AAT CCC CTG AGC GTC AGT GTY AAC CC	This study This study
AOB1150/ Nso1225	β-AOB 16S rDNA	GCG CAA CCC TTG TCA YTA ATT G CGC CAT TGT ATT ACG TGT GA	This study (Mobarry et al. 1997)
amoA-1F/ amoA-2R	*amoA* gene	GGG GTT TCT ACT GGT GGT CCC CTC KGS AAA GCC TTC TTC	(Rotthauwe, Witzel, and Liesack 1997)
amoA-Ns1	*Nitrosospira amoA*	CCG ACS CAC CTG CCG CTG G	This study
amoA-Nm3	*Nitrosomonas amoA*	TGT CGA TGG CTG AYT ACA TGG G	This study
amoA-Nm4	*Nm. oligotropha amoA*	ATC ATG TTG CTG ACC GGT AAC TGG C	This study

One potential drawback of using *amoA* for quantitation of AOB is the different copy numbers of this gene in the AOB genera, with *Nitrosomonas* species apparently having two copies and *Nitrosospira* strains having three copies (Klotz, and Norton 1998). Therefore, transforming real-time PCR data on *amoA* copy numbers to an estimate of AOB cell numbers would require the differentiation between these two genera. To discriminate between these two templates, we initially attempted to develop an approach that combines the TaqMan® and intercalating dye real-time PCR techniques. This combined strategy entailed selective amplification of *amoA* using primers amoA-1F and amoA-2R coupled with: (1) general detection of all *amoA* amplicons using the intercalating dye SYBR Green, and (2) specific detection of *Nitrosospira*-derived *amoA* using a *Nitrosospira*-targeted TaqMan probe. The concentration of *Nitrosomonas*-derived *amoA* would then be determined by difference between the intercalating dye and TaqMan results. Probe amoA-Ns1 (Table 3.4) was designed to target all known *Nitrosospira amoA* sequences, and has six mismatches with known *Nitrosomonas* sequences. The reason for using an intercalating dye to quantify *Nitrosomonas amoA* is the diversity of sequences in this genus preventing the design of a specific probe.

Following this attempt at a coupled TaqMan and SYBR Green protocol to differentiate *Nitrosomonas*- and *Nitrosospira*-derived PCR products, a multiplex TaqMan approach was developed due to problems with primer-dimer interference in the SYBR Green signal. This strategy entailed selective amplification of *amoA* using primers amoA-1F and amoA-2R coupled with the specific detection of *Nitrosospira*-derived *amoA* using a Hex-labeled probe (amoA-Ns1) and *Nitrosomonas*-derived *amoA* using a 6-FAM-labeled probe. Because of the sequence diversity within the *Nitrosomonas* genus, it was not possible to design a single probe that uniquely complemented all *Nitrosomonas* sequences. Therefore, two probes (amoA-Nm3 and amoA-Nm4) were used separately. The specificity of probe amoA-Nm3 was broadened by including one degenerate position (i.e., a location at which target sequences have multiple nucleotides), and the known marine *Nitrosomonas* sequences were not targeted. This probe included six mismatches with known *Nitrosospira* sequences. Probe amoA-Nm4 was designed to specifically target *Nitrosomonas oligotropha*, which was expected in these systems based on previous characterization research (Regan, Harrington, and Noguera 2002; Regan et al. 2003). Primer and probe sequences are listed in Table 3.4. All TaqMan probes were 3'-labeled with BH1 (Black Hole 1) as the quencher dye because of its broad quenching spectrum and lower signal to noise ratio than that of other quenching dyes. All probes used in this study were synthesized by Sigma Genosys (Woodlands, TX).

PCR Protocol

For the 16S rDNA-targeted PCR, reactions were performed in 50 μL solutions containing 0.1 mg/mL bovine serum albumin (BSA), 0.2 μM forward and reverse primers, a range of 1.5 to 3.0 mM $MgCl_2$, and 0.2 mM dNTPs. BSA was added to reduce inhibition by humic substances (Kreader 1996). After a 5 min hot start at 80°C, 2.5 U Taq polymerase (Promega, Madison, WI) was added. Using a temperature gradient thermocycler (iCycler, Bio-Rad, Hercules, CA), a range of annealing temperatures was evaluated for each primer pair to examine reaction specificity. PCR products were inspected using agarose gel electrophoresis, and digital images were acquired using a ChemiDoc imager and Quantity One software (Bio-Rad, Hercules, CA).

For *amoA*-targeted real-time PCR, the total reaction volumes were 25 μL consisting of 12.5 μL Bio-Rad iQ Supermix (2x mix containing 100 mM KCl, 40 mM Tris-HCl, pH 8.4, 0.4 mM each dNTP, 50 U/mL iTaq DNA polymerase, 6 mM $MgCl_2$, stabilizers), 5 uL DNA, 1 uL each of forward and reverse primers (final concentration 200 nM), 0.5 uL TaqMan probe (final concentration 100 nM), 2.5 uL BSA (final concentration 0.1 mg/mL), and 2.5 uL water. The thermal profile included an initial denaturation step at 95°C for 30 seconds, and then 50 cycles of denaturation at 95°C for 40 seconds, annealing at 58°C for 1 minute, and elongation at 72°C for 45 seconds. Cycling was completed with a final elongation step at 72°C for 1 minute. Amplifications were performed with a Bio-Rad iCycler iQ^{TM} real-time detection system (Bio-Rad, Hercules, CA) using 96-well reaction plates (Bio-Rad) sealed with optically clear sealing film (Bio-Rad). PCR samples and controls were prepared in triplicate. A negative control with no template DNA was run in each reaction. To test for the presence of PCR inhibitors, positive control DNA from the pure cultures was spiked into extracted DNA from environmental samples from each site. Samples with and without the addition of positive control DNA were compared, and the percent recovery of the added DNA was calculated. Data analysis was carried out with iCycler iQ^{TM} software (version 3.1), and the size of PCR products was verified with agarose gel electrophoresis.

Standard Curves

Standard curves for probes amoA-Ns1 and amoA-Nm3 were initially constructed on the basis of cell numbers, using serial dilutions of genomic DNA extracted from the AOB pure cultures. Cell concentrations were determined by DAPI staining as detailed below. DNA was extracted from these known cell concentrations and serially diluted, and *amoA* was amplified and detected using the primers and probes listed in Table 3.4. The Ct for each standard was plotted against the log of initial cell number to construct the standard curve used to quantify AOB. To compare PCR efficiencies and detection sensitivities among different experiments, slopes of the standard curves were calculated by linear regression analysis with the iCycler iQ^{TM} software. Amplification efficiency (E) was calculated from the slope using the following equation:

$$E = \left(10^{-1/slope}\right) - 1 \quad (3.1)$$

With the specific targeting of *Nitrosomonas oligotropha* by probe amoA-Nm4 and the lack of an isolate of this species, standard curves were also generated for each probe using serial dilutions of cloned *amoA* fragments derived from each target group. For probes amoA-Ns1 and amoA-Nm3, *amoA* fragments were retrieved from the pure AOB cultures. For amoA-Nm4, clones of *amoA* were constructed from PCR products retrieved with drinking water samples collected July 9, 2004 at ACWD. PCR products generated with the *amoA*-1F and *amoA*-2R primer set were purified using QIAquick PCR Purification Kit (Qiagen, Valencia, CA) according to the manufacturer's instructions. These products were cloned into vector pCR®2.1 using the TOPO TA cloning kit (Invitrogen Corp., San Diego, CA). The ligation reaction included 4 µL of PCR product, 1 µL of salt solution, and 1 µL of vector. After incubation for 5 minutes at room temperature, 2 µL of the solution was transformed into *Escherichia coli* One Shot TOP 10 chemically competent cells (Invitrogen Corp., San Diego, CA). The cells were heat shocked at 42°C for 30 seconds, and then immediately transferred to ice. After the addition of 250 µL of S.O.C medium, these transformed cells were incubated at 37°C for 1 hour with horizontal shaking, and then 100 µL of each transformation was spread on a prewarmed selective Luria Bertani (LB) plate including ampicillin and X-gal (5-bromo-4-chloro-3-indolyl-beta-D-galactopyranoside). Blue and white screening was performed after 14 hours. Five white colonies from each plate were cultured overnight in LB medium containing 50 ug/mL ampicillin. Plasmid DNA was isolated using the QIAprep spin miniprep kit (Qiagen, Valencia, CA) following the manufacturer's instructions. Cloned inserts were sequenced at The Pennsylvania State University's Nucleic Acids Facility and analyzed using BLAST to confirm their identity.

Serial 10-fold dilutions of this purified plasmid DNA were used for standard curve generation. The amount and purity of plasmid DNA were determined spectrophotometrically with a Shimadzu UV-1601 spectrophotometer by measuring the optical densities at 260 nm (A_{260}) and 280 nm (A_{280}). The DNA used for standard curve generation possessed an A_{260}/A_{280} ratio higher than 1.8 (Sambrook, Fritsch, and Maniatis 1989). With a known basepair length of the vector plus PCR insert (*BP*), the copy number of *amoA* was calculated from the concentration of extracted plasmid DNA (*X*, g/µL) using the following formula:

$$\frac{copies}{\mu L} = 6.022 \times 10^{23}\left(\frac{X}{BP \times 660}\right) \quad (3.2)$$

where 660 is the average molecular weight of a nucleotide basepair. Gene copy number data from environmental samples was then converted to estimates of AOB cell numbers with the assumptions that *Nitrosospira* genomes contain three copies of *amoA* and *Nitrosomonas* genomes have two copies.

CELL COUNTING

Pure culture cell densities were determined using DAPI staining and epifluorescence microscopy. A 100 to 200 uL aliquot of *Nitrosomonas europaea* or *Nitrosospira multiformis* culture was mixed with 5 mL of DAPI solution (2 µg/mL) and incubated in the dark for 5 min. This suspension was then filtered through a 25 mm diameter, 0.22 µm black polycarbonate filter

(Osmonics Inc., Minnetonka, MN) and washed with 10 to 20 mL of distilled water. The filter was then placed on a glass microscope slide and mounted in DABCO antifade solution. Images were captured using a 63x objective lens and a Zeiss Axiophot epifluorescent microscope (ZEISS, Germany), and cells were counted from these images. Total cells were enumerated and averaged from 20 different images, and the cell density was calculated using the following formula:

$$\frac{Cells}{mL} = \frac{Count \times \left(\frac{A_1}{A_2}\right)}{Volume} \tag{3.3}$$

where A_1 (area of filter) = 2.269 cm^2 and A_2 (area of image) = 15,273 μm^2.

MOST-PROBABLE NUMBER (MPN)

AOB concentrations in water samples were also enumerated by a five-tube MPN procedure with four tenfold serial dilutions using Soriano and Walker medium (Soriano, and Walker 1968). The medium is composed of the following ingredients per L of distilled water: 0.5 g of $(NH_4)_2SO_4$, 0.2 g of KH_2PO_4, 0.04 g of $CaCl_2 \cdot 2H_20$, 0.04 g of $MgSO_4 \cdot 7H_2O$, and 0.5 mg of Fe-EDTA. The pH of the medium was adjusted to 8.0 with 5% K_2CO_3 solution. Each MPN tube contained 9 mL of medium and was inoculated with 1 mL of sample or the preceding dilution. After inoculation, tubes were incubated in the dark for 28 days at 30 °C. At the end of the incubation period, MPN tubes were tested for the presence of nitrite using sulfanilic acid and N,N-dimethyl-α-naphthylamine (APHA, AWWA, and WEF 1995). The production of a pink color within 5 minutes indicated the tube was positive for the presence of AOB (Wolfe et al. 1990). The detection limit of the assay was 20 AOB per 100 mL.

CHAPTER 4
RESULTS AND DISCUSSION – FLUORESCENT IN SITU HYBRIDIZATION STUDIES

MICROSCOPY-BASED ANALYSES

While the goal of the FISH effort was to develop a FCM-based protocol to rapidly enumerate AOB, epifluorescent microscopy was first used to evaluate several strategies for enhancing target cell discrimination. This section presents the microscopy-based results of these FISH modifications.

Single-Probe Hybridizations

The initial FISH effort involved characterizing the signal intensities from single-probe hybridizations. Probes Nso1225 and AOB770 were used to target AOB, EUB338 as a positive-control probe, nonEUB338 as a negative-control probe to measure nonspecific binding, and a no probe control to measure cell autofluorescence. Probe sequences and associated FISH conditions are presented in Table 3.3. AOB770, which was designed in this project, was evaluated over a range of hybridization and washing conditions (variable formamide and salt concentrations, respectively) and found to confer a uniform fluorescent intensity to target AOB cells up to 50% formamide, while not yielding an elevated fluorescence from non-target cells. Therefore, the conditions used for this probe were matched to Nso1225 for probe compatibility.

As expected, both Nso1225 and AOB770 successfully discriminated *Nm. europaea* and *Ns. multiformis* from the negative control cultures. With the non-target control cells, the fluorescent intensity was the same as autofluorescence with these AOB-specific probes. EUB338 hybridization imparted fluorescence to all cell types, indicating that cell permeability and ribosome abundance were sufficient in the negative control cultures. Mixtures of the target and non-target pure cultures were analyzed to assess the ability of detecting target AOB within a background of non-target bacteria, as would be encountered in drinking water. In these mixed cell preparations, all cells were visualized with the EUB338 probe. However, only AOB hybridized with the AOB-specific probes Nso1225 and AOB770.

Dual-Probe Hybridizations

Though the AOB-specific probes were successful at detecting AOB pure cultures, the signal of cells from environmental samples might not be as bright due to the lower ribosome content of cells growing under the oligotrophic conditions encountered in distribution systems. In addition, drinking water samples often contain materials that contribute background fluorescence, which would necessitate a strong probe-conferred signal to allow unequivocal AOB enumeration. To increase the FISH signal, Nso1225 and AOB770 were hybridized simultaneously. These probes target different locations on the rRNA and have compatible hybridization conditions, allowing the potential for dual-labeling of target rRNA. Figure 4.1 shows the average AOB cell intensities of single- and combined-probe hybridizations, background subtracted using the intensities of no-probe controls. The intensities of the

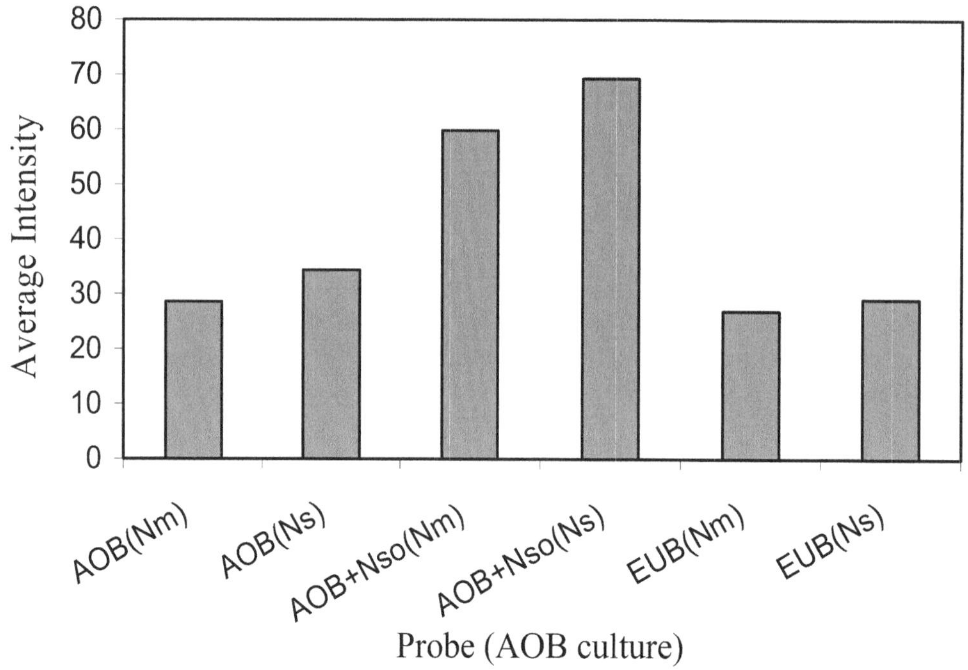

Figure 4.1 Average AOB cell intensities for single- and dual-probe hybridizations.

combined probes were double those of single probes for both *Nitrosomonas* and *Nitrosospira* control cultures, suggesting that the two probes target approximately equally accessible regions of the 16S rRNA. These data show that the use of combined probes can increase the intensities to assist in detecting target cells with a low rRNA content and above elevated background fluorescence.

Tyramide Signal Amplification

Another strategy that was tested to enhance target-cell fluorescent intensity was TSA using fluorescein-conjugated tyramide combined with HRP-labeled probes targeting AOB (Nso1225) and the *Bacteria* domain (EUB338). Figure 4.2 shows representative images of TSA FISH compared with conventional single-probe FISH. TSA yielded a considerable increase in cell fluorescence. Average cell intensities, presented in Table 4.1, were measured from images captured using identical exposure times. The signal intensities from TSA FISH were 3.5 to 4.5 times brighter than those of traditional single-probe FISH, and approximately twice as bright as the dual label results. This was achieved without an increase in background fluorescence associated with non-target cells. The total sample preparation time for TSA FISH was 1.5 times longer than that of traditional FISH due to additional steps in the protocol such as lysozyme treatment to make the cells permeable to HRP and tyramide activation. While the observed intensity increases appear promising for improving the discrimination of AOB from non-target cells, the major limitation of this approach is the cost of the fluorescein-conjugated tyramide. Slide preparations for microscopy use small volumes of this reagent, but larger volumes would be required for flow-cytometric preparations. Therefore, this effort was not pursued further for this project.

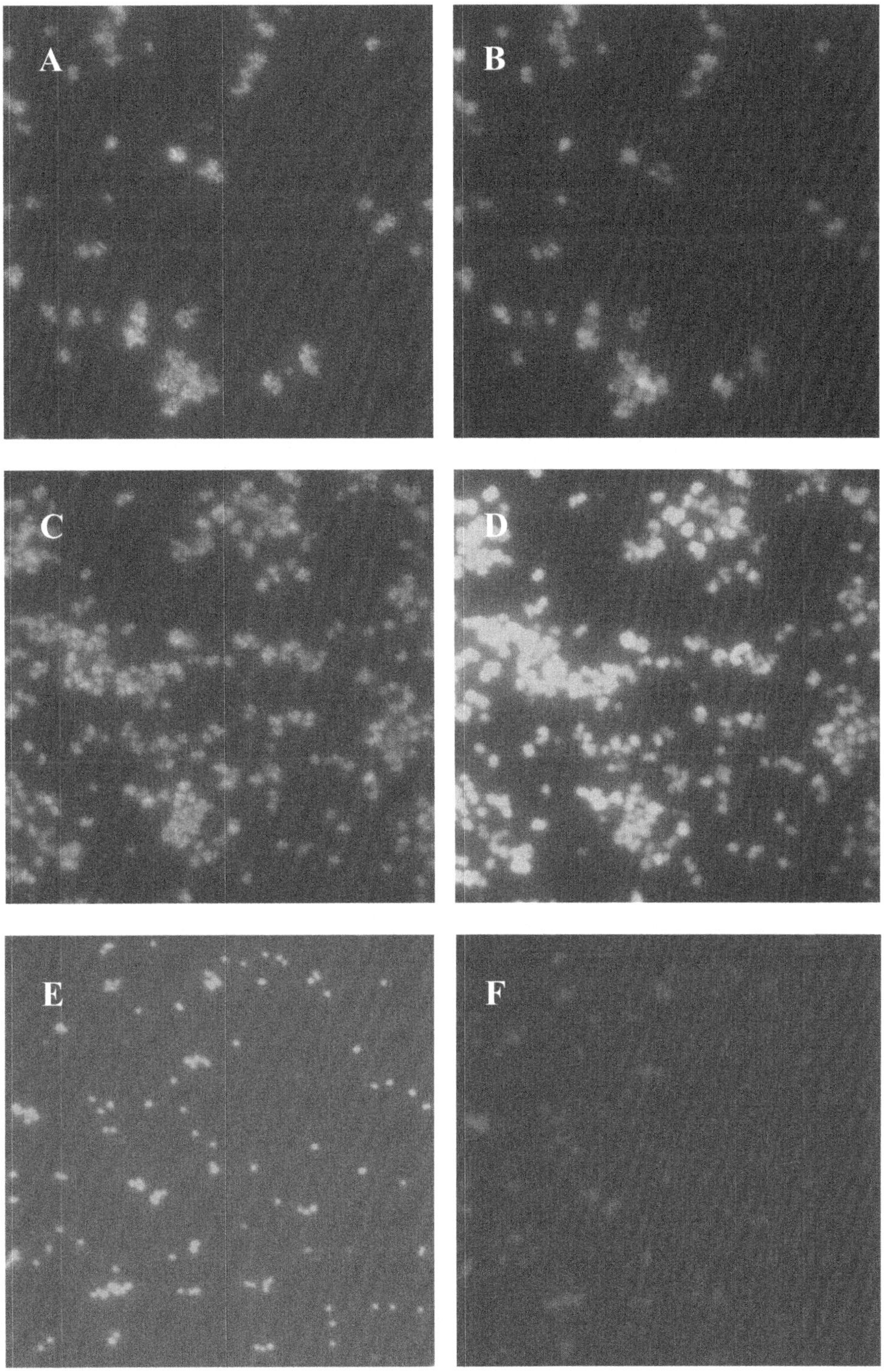

Figure 4.2 Images of traditional FISH and TSA FISH, with DAPI staining on the left panels and accompanying FISH images on the right panels. A-B: *Ns. multiformis* with FISH using Nso1225, C-D: *Ns. multiformis* with TSA FISH using Nso1225, and E-F: *Nm. europaea* and *E. coli* with TSA FISH without probe.

Table 4.1
Comparison of cell intensities using conventional and tyramide signal amplificiation FISH

	Nm. europaea			*Ns. multiformis*		
	Conventional	TSA	Increase (%)	Conventional	TSA	Increase (%)
Nso1225	27.5	107	289	36.6	167	356
EUB338	29.5	125	324	44.4	157	254

FLOW CYTOMETRIC ANALYSES

Traditional FISH

Target and non-target cells were hybridized with the AOB-specific probes Nso1225 and AOB770, the positive control probe EUB338, and the negative controls with nonEUB338 probe and no probe. These preparations were then detected by FCM. In the first step of FCM data analysis, the dot plots based on forward scatter (FS) and side scatter (SS) were gated into regions to define cells, as depicted by region A in Figure 4.3. Fluorescence histograms of bacterial sized particles were then produced from these defined regions. For example, Figure 4.4 shows the histograms (designated FITC) generated from the *Nm. europaea* control culture, with the no-probe control (panel A), the non-EUB338 probe for non-specific binding (panel B), the AOB-specific probe AOB770 (panel C), and combined probes Nso1225 and AOB770 (panel D). These histogram data were converted using the program MFI into linearized data, and the linearized mean intensity values are presented in Table 4.2 for *Nm. europaea, Ns. multiformis,* and *E. coli*.

With *Nm. europaea* and *Ns. multiformis,* the linearized mean relative fluorescent intensities from dual-labeling with Nso1225 and AOB770 were seven to eight times larger than single-probe labeling with either AOB-targeted probe. Probe EUB338 gave a comparable mean cell intensity with *Nm. europaea* as the AOB-targeted probes, but was only about half the signal

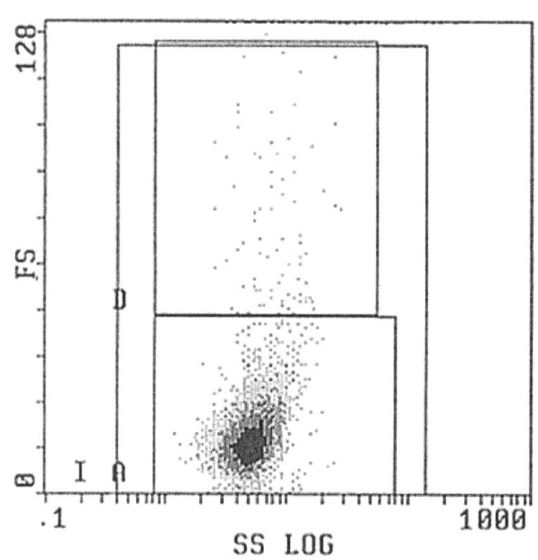

Figure 4.3 Dot plot of forward scatter (FS) versus sidescatter (SS) with *Nm. europaea*

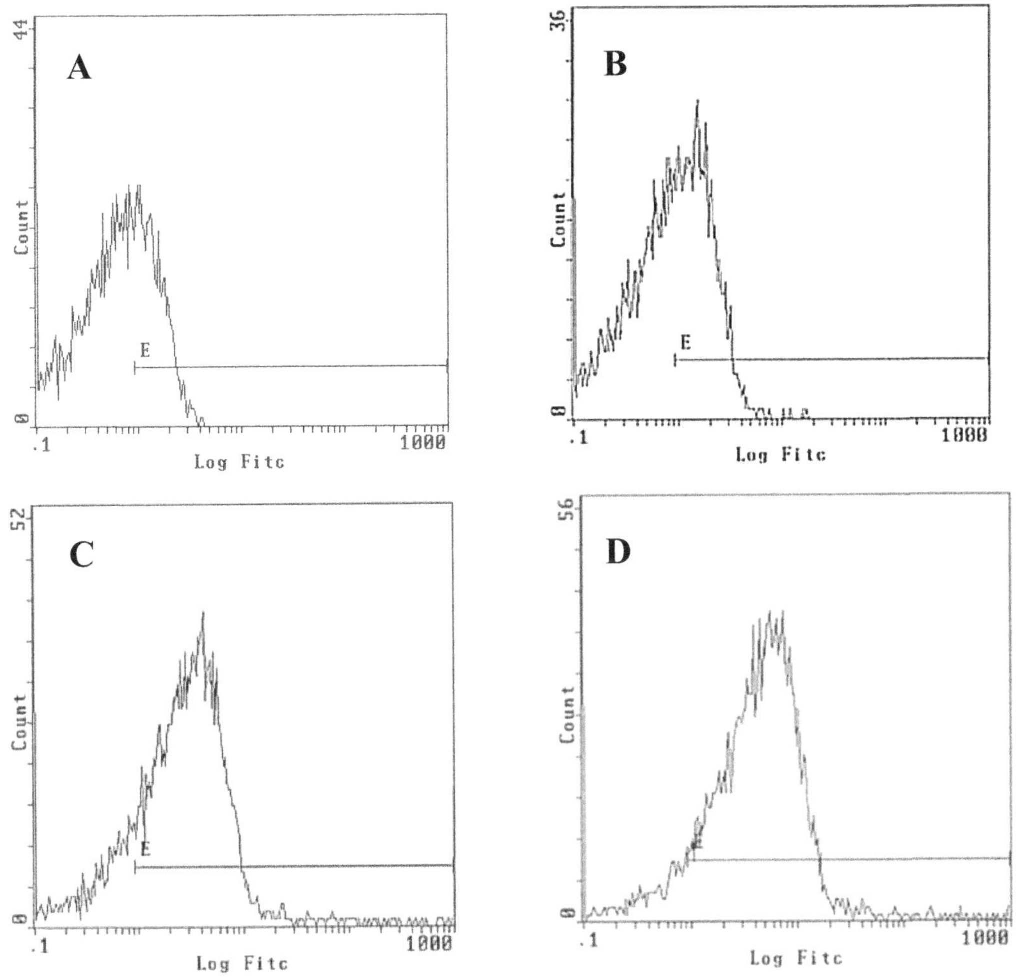

Figure 4.4 Histograms of fluorescence with *Nm. europaea*. A: no probe control, B: nonEUB338 probe, C: AOB770 probe, and D: combined Nso1225 and AOB770 probes

Table 4.2
Mean intensities from FCM analyses

	Nm. europaea		*Ns. multiformis*		*E. coli*	
	Log	Linear	Log	Linear	Log	Linear
No probe	0.697	4.98	0.754	5.68	0.738	5.47
NonEUB338	1.41	25.70	1.57	37.15	1	10.0
EUB338	2.03	107.15	1.32	20.89	2.06	114.82
Nso1225	2.09	123.03	1.76	57.54	0.876	7.52
AOB770	2.12	131.83	1.74	54.95	1	10.0
Nso1225 +AOB770	2.98	954.99	2.63	426.58	1.14	13.80

intensity for *Ns. multiformis*, less than the nonEUB338 control. It is not clear why EUB338 did not show evidence of hybridizing with *Ns. multiformis* in this set of experiments. With the exception of the negative controls, *Ns. multiformis* showed a lower average intensity in these analyses than *Nm. europaea*. This is inconsistent with the microscopy-based analyses, in which *Ns. multiformis* generally had slightly higher cell intensity than *Nm. europaea*. This may simply be due to the growth rate of the respective cells at the time of cell harvesting. The non-EUB338 data show that there was some non-specific binding of probes, but the resultant fluorescence was significantly below probe-conferred intensities. The results with *E. coli* showed a positive response with EUB338, showing permeability and adequate ribosome content of the cells. The AOB-specific probes, hybridized either separately or simultaneously, did not show an elevated response with these non-target cells. These results suggest that the combined AOB-targeted probes could be used to differentiate AOB and non-target cells. While the probe indeed conferred an increase in cell fluorescence relative to non-specific binding, it remains untested whether this relative increase would be sufficient to enumerate AOB cells with low rRNA content amid a background of non-target cells and debris.

The primary purpose of FCM in this study was the rapid quantitative analysis of AOB cells. To investigate whether the FCM-FISH protocol results in a significant loss of bacteria during the sequential filtering and resuspension steps, pure culture enumeration with FCM was performed in parallel with direct microscopic quantification. FCM flowrates were calculated using a known concentration of fluorescent beads. The microscopic count of cells was performed by filtering a culture, DAPI staining, and direct counting with the microscope. The cell count results using both methods are displayed in Table 4.3. These results show that the cell numbers from FCM analyses were less than that by the microscopic method. For *Nm. europaea*, the FCM cell counts were 59.4% to 82.3% compared with the microscopic direct counts. The *Ns. multiformis* cell numbers from FCM were only 6.7% to 14.1% of those from microscopic counts. One possible explanation for the apparently greater cell loss for *Ns. multiformis* might be due to errantly high counts from the microscope due to the partitioned shape of *Ns. multiformis* cells. For both AOB cultures, the cell number using dual labeling was significantly higher than that of single probes, suggesting that the use of the combined probes can provide a more accurate way of counting the suspended cells using FCM due to increased signal intensities. Improvements may be necessary in the multiple transfer steps of the FISH-FCM protocol to reduce the number of cells lost during preparation.

Table 4.3
Quantitative analysis of target cell recovery

		Nm. europaea (cells/ml)	*Ns. multiformis* (cells/ml)
Microscopy	DAPI stain	7.81×10^7	5.93×10^6
Flow cytometry	Nso1225	4.64×10^7	4.57×10^5
	AOB770	5.23×10^7	4.45×10^5
	EUB338	5.12×10^7	3.98×10^5
	Nso1225 + AOB770	6.43×10^7	8.37×10^5

TSA FISH

The TSA method was also applied to FCM FISH analysis. The protocol for FCM was developed in this study by combining the protocols for FCM analysis of traditional FISH with that of microscopy-based TSA FISH. The analysis included two HRP-labeled probes, Nso1225 and EUB338, and no-probe negative controls.

The experiments with target and non-target cells showed that signal intensities from TSA were much brighter than those from single-fluorochrome FISH probes. The raw log-scale intensities from traditional and TSA FISH by FCM are summarized in Table 4.4. The intensities with the non-target cultures *E. coli* and *R. basilensis* using the Nso1225 probe were distinctly lower than those from the EUB338 probe. In addition, the ratios (traditional FISH:TSA FISH) were calculated, following conversion from log to linear values. The ratios indicate that the intensities by TSA FISH were increased between 30- to 100-fold relative to those of traditional FISH. Figure 4.5 shows the histograms from FCM of the no-probe control and using HRP-labeled Nso1225 probe with *Nm. europaea*, illustrating the large increase in intensity due to probe hybridization and tyramide amplification. The intensities using TSA flow cytometric FISH were seven- to eight-fold brighter than those of traditional combined probes and 30- to 100-fold for traditional single probe. Based on these results, the application of TSA might be effective to detect and enumerate AOB. However, because only pure cultures were analyzed in these FCM experiments, the results of this study do not allow an evaluation of the ability of FCM to differentiate between target bacteria and background events that might be present in real drinking water samples. Given the costs of HRP-labeled probes and TSA reagents, this protocol was not pursued further. The development a cost-effective FCM-based assay would require modifications to the protocol that reduce the volumes of these components.

Table 4.4
Comparison of the log intensities by FCM FISH

		Nm. europaea	*Ns. multiformis*	*E. coli*	*R. basilensis*
No probe	Traditional	0.376	0.175	0.104	0.137
	TSA	1.15	1.18	1.01	1.13
	(TSA/Trad.)*	5.94	10.1	8.05	9.84
Nso1225	Traditional	2.75	3.73	1.08	1.15
	TSA	4.57	5.67	1.13	1.38
	(TSA/Trad.)*	66.1	87.1	1.12	1.70
EUB338	Traditional	2.38	3.24	2.98	2.67
	TSA	4.17	5.1	4.99	4.12
	(TSA/Trad.)*	61.7	72.4	102.3	28.2

* linearized intensity by TSA / linearized intensity by traditional FISH

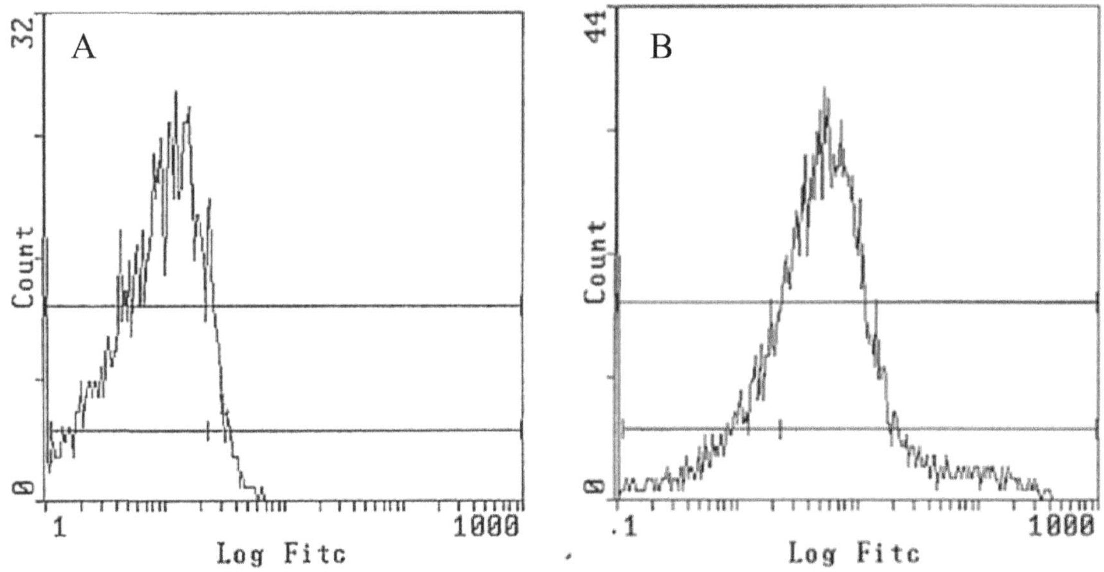

Figure 4.5 Fluorescence histograms from FCM using TSA FISH with *Nm. europaea*, A: no probe, and B: Nso1225 probe

CHAPTER 5
RESULTS AND DISCUSSION – REAL-TIME POLYMERASE CHAIN REACTION STUDIES

PROTOCOL DEVELOPMENT

16S rDNA-Targeted PCR

The initial PCR effort focused on targeting the 16S rRNA gene (16S rDNA) because it is present on the chromosome in only a single copy in all characterized AOB, thereby allowing a direct connection between gene copy number and cell number. With the primer pairs listed in Table 3.4 and pure cultures of target and non-target cells, primer specificity was tested using conventional PCR assays. Both the AOB 620F and AOB 770R primer pair and the AOB 1150F and Nso1225R primer pair produced PCR products of the expected size with positive control DNA. However, the negative control samples were also amplified with each primer pair, even though the sequences of the negative control DNA had at least two nucleotide mismatches with each primer pair. These nonspecific PCR products could not be eliminated without observing a decreased yield of AOB-derived products.

This problem of nonspecific amplification products has been encountered by other researchers attempting to retrieve the molecular diversity of β-subgroup AOB. In an investigation of AOB in soil and marine environments, Stephen et al. (1996) reported the coamplification of environmental sequences closely related to sequences of non-nitrifying members of β-subgroup of *Proteobacteria*, including *Comamonas testosterone*, *Sphaerotilus natans*, and *Rubrivivax gelatinosus*. Because the 16S rRNA gene is present in all prokaryotes, primers targeting 16S rRNA genes can cross-react with members of other phylogenetic and physiological groups. Therefore, instead of using a 16S rDNA-based primer pair, all subsequent PCR work on this project targeted the functional gene *amoA*.

Combined TaqMan and SYBR Green Approach

The initial strategy of AOB enumeration by real-time PCR involved the amplification of *amoA* followed by TaqMan probe detection of *Nitrosospira*-derived products and SYBR Green detection of all products. *Nitrosomonas* concentration would then be determined by difference between the SYBR Green and TaqMan concentrations. The reason for differentiating *Nitrosomonas* and *Nitrosospira* products is that these genera appear to have a different copy number of the *amoA* gene. Of those species that have been studied, *Nitrosomonas* have two copies and *Nitrosospira* have three copies of the *amoA* gene (Klotz, and Norton 1998). The real-time PCR TaqMan and SYBR Green results showed a good linear correlation between threshold cycle (C_T) and template DNA concentration for *Nitrosospira* and *Nitrosomonas* templates (Figures 5.1 and 5.2, respectively). Also, the TaqMan probe amoA-Ns1 (Table 3.4) did not hybridize with *Nitrosomonas* template (data not shown), indicating the desired probe specificity. However, the SYBR Green detection of *Nitrosomonas* amplicons was not very sensitive, only giving a linear response over four orders of magnitude (Figure 5.2). Moreover, the melting curves for the *Nitrosomonas* SYBR Green samples indicated the formation of primer dimers by the presence of peaks at temperatures below the melting temperature of the amplification products (Figure 5.3). Primer dimers interfere with accurate quantitation of template concentration by intercalating dyes, which do not discriminate PCR products. Therefore, this hybrid strategy of TaqMan and SYBR Green detection was abandoned for the development of distinct TaqMan probes for each target AOB group.

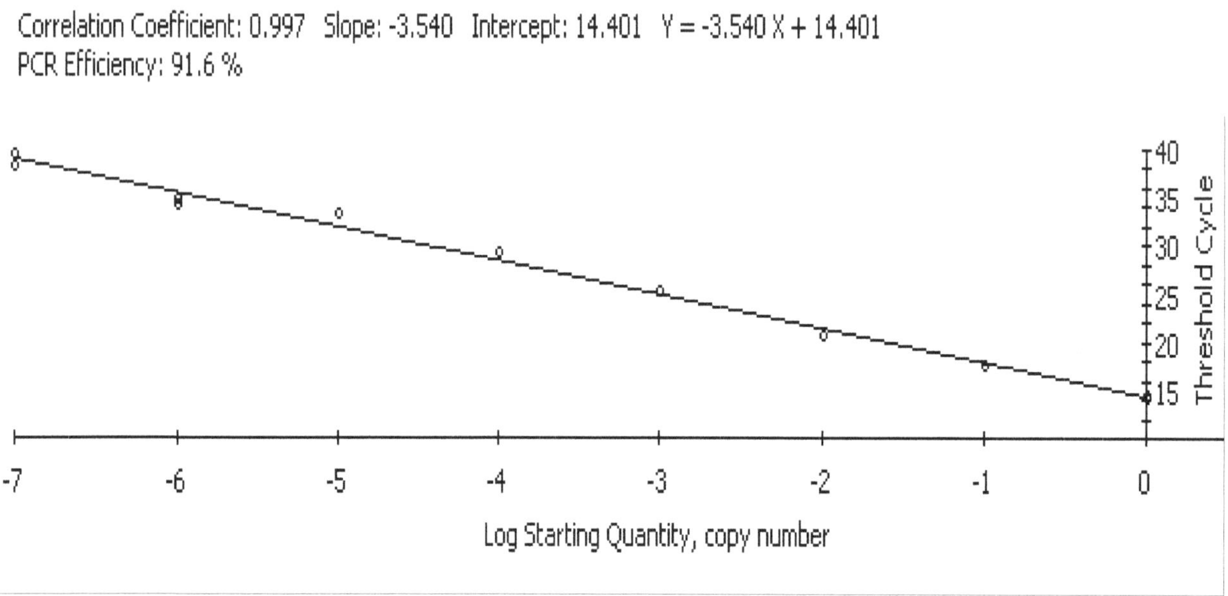

Figure 5.1 Standard curve for *Nitrosospira multiformis* using serially diluted genomic DNA and TaqMan probe amoA-Ns1

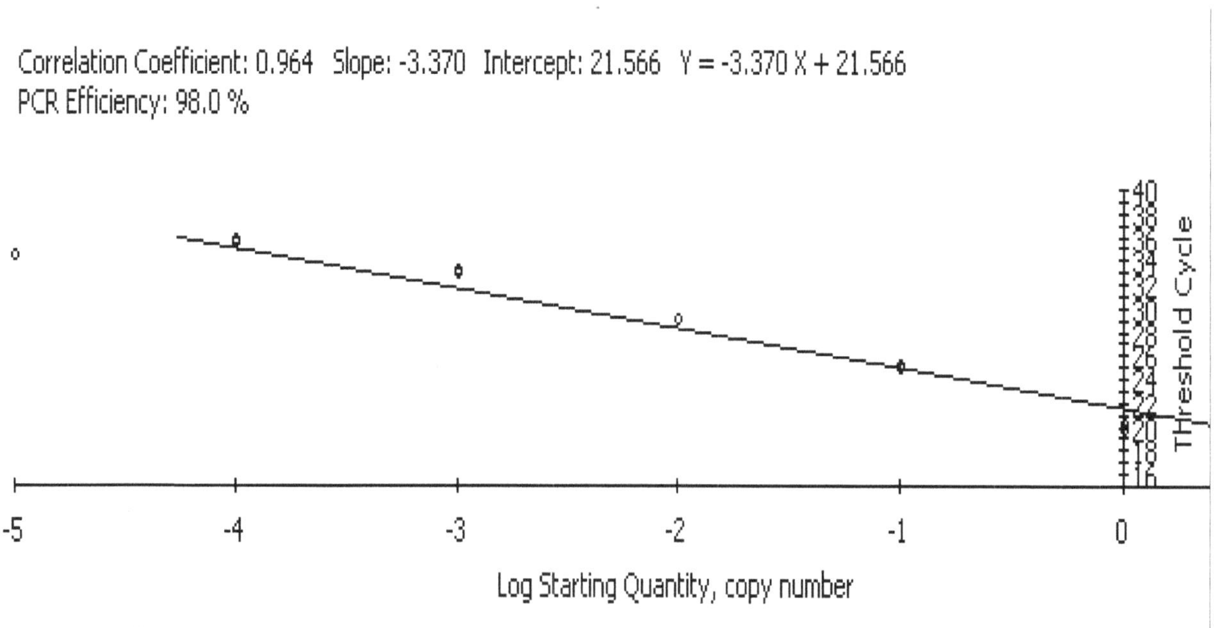

Figure 5.2 Standard curve for *Nitrosomonas europaea* using serially diluted genomic DNA and SYBR Green detection

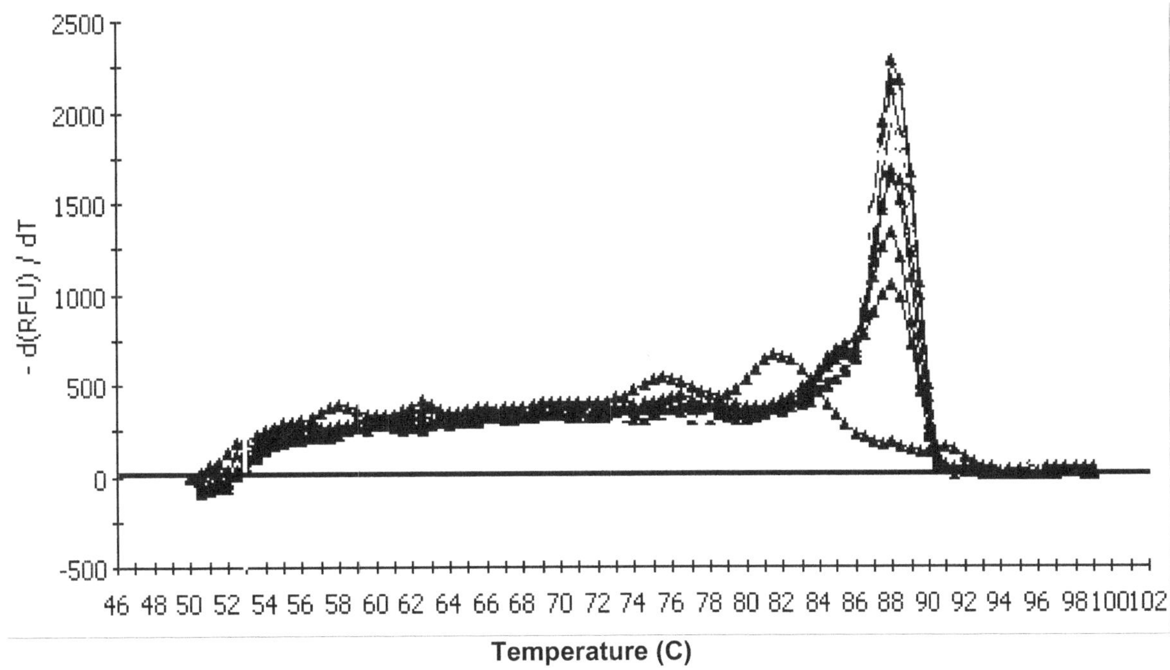

Figure 5.3 Melting curves for *Nitrosomonas* using serially diluted genomic DNA and SYBR Green detection

TaqMan Approach

To overcome the problem of primer dimer interference, the strategy was modified to include separate TaqMan probes for *Nitrosospira* and *Nitrosomonas*, each labeled with a distinct fluorochrome. Because of the sequence diversity within the *Nitrosomonas* genus, it was not possible to design a single probe that uniquely complemented all *Nitrosomonas* sequences. Therefore, two *Nitrosomonas* target groups were identified, non-marine species and *Nitrosomonas oligotropha*. To broaden the specificity of probes targeting this genus, degenerate positions (i.e., locations at which target sequences have multiple nucleotides) were included in the probe designs. The first two designs of *Nitrosomonas*-targeted probes, amoA-Nm1 and amoA-Nm2, had linear standard curves but poor PCR efficiencies and linear ranges. For example, the results of probe amoA-Nm1 (Figure 5.4) showed an efficiency of only 77.4%. The third attempt at a *Nitrosomonas*-targeted TaqMan probe was successful (Figure 5.5), with probe amoA-Nm3 (Table 3.4) showing a good standard curve with *Nitrosomonas europaea* and no hybridization to *Nitrosospira multiformis* amplification products. A third TaqMan probe was designed to target the *amoA* sequence of *Nm. oligotropha* (GenBank accession number AF272406). This probe, amoA-Nm4, had at least six and five degenerate positions with the AOB sequences belonging to the *Nm. europaea* cluster and the *Nitrosospira* cluster, respectively.

Given the lack of a pure *Nm. oligotropha* culture for use as a control, *amoA* fragments were amplified, cloned, and sequenced from a chloraminated reservoir sample as described in Chapter 3. BLAST analysis of the sequences and alignment with other known *amoA* sequences confirmed that they were derived from *Nm. oligotropha*-like AOB, with 95 to 96% identity to the *amoA* gene from *Nm. oligotropha* AF272406 (Figure 5.6). This is consistent with previous AOB characterizations from chloraminated systems (Regan, Harrington, and Noguera 2002; Regan et al. 2003), which reported that all 16S rDNA clone sequences retrieved from a pilot-scale and several full-scale systems were strongly affiliated with the *Nm. oligotropha* cluster. The amoA-Nm4 probe sequence was a perfect match to each of the reservoir-derived cloned sequences. Likewise, *amoA* fragments were also cloned from the *Nitrosomonas europaea* and *Nitrosospira multiformis* pure cultures. All subsequent probe specificity testing and standard curve development was performed with purified plasmid from these three clones. TaqMan probe amoA-Nm4 was shown to be very specific to the target *Nm. oligotropha* sequence, and no cross-reactivity was observed between this probe and the other two TaqMan probes, amoA-Nm3 and amoA-Ns.

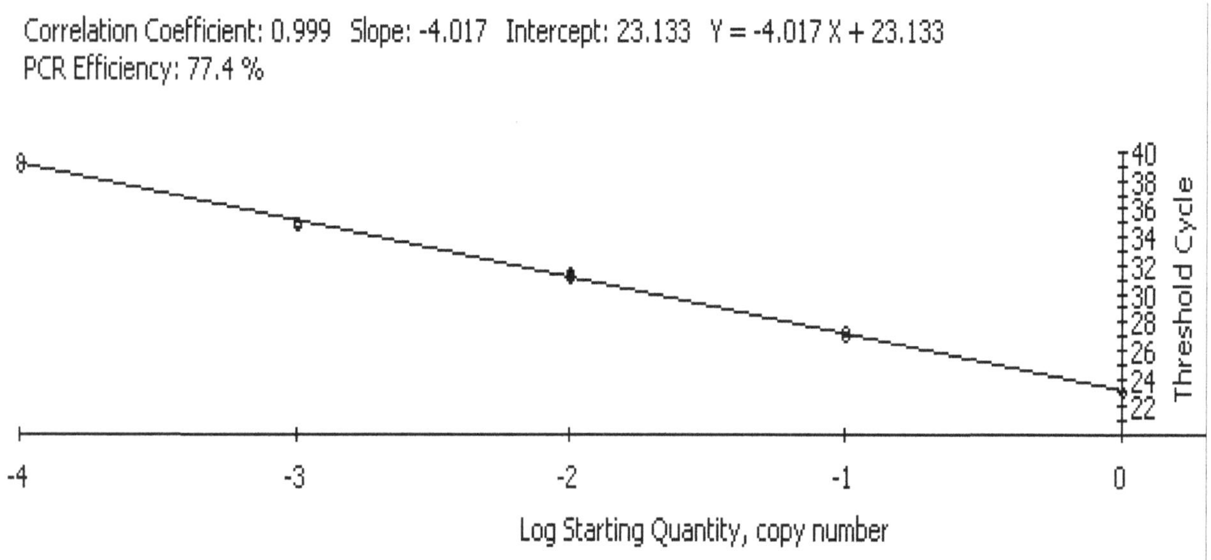

Figure 5.4 Standard curve for *Nitrosomonas* using TaqMan probe amoA-Nm1

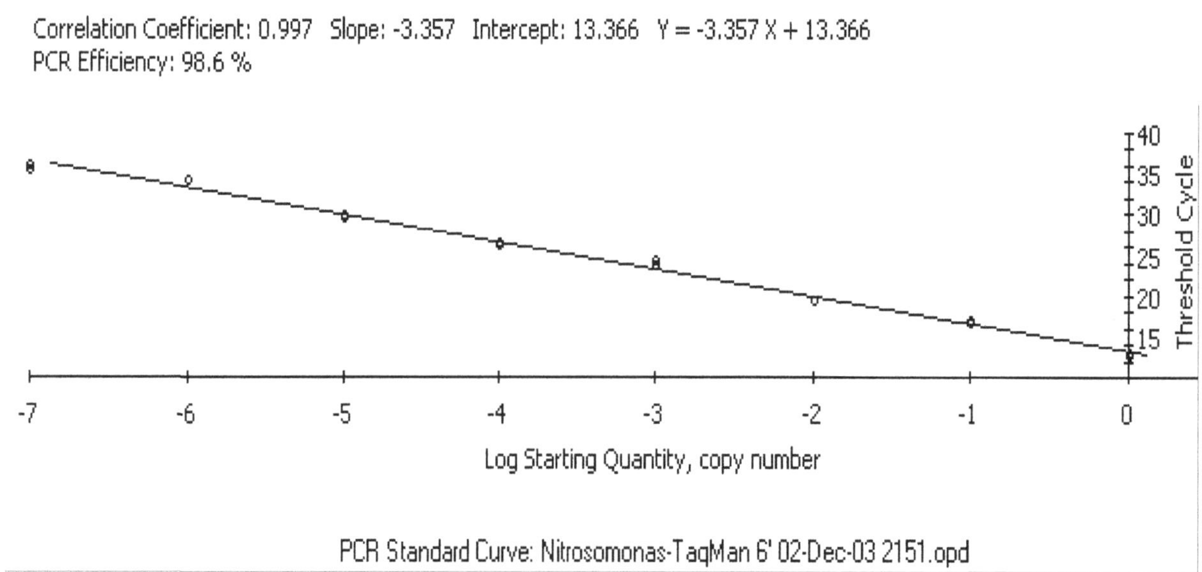

Figure 5.5 Standard curve for *Nitrosomonas* using TaqMan probe amoA-Nm3

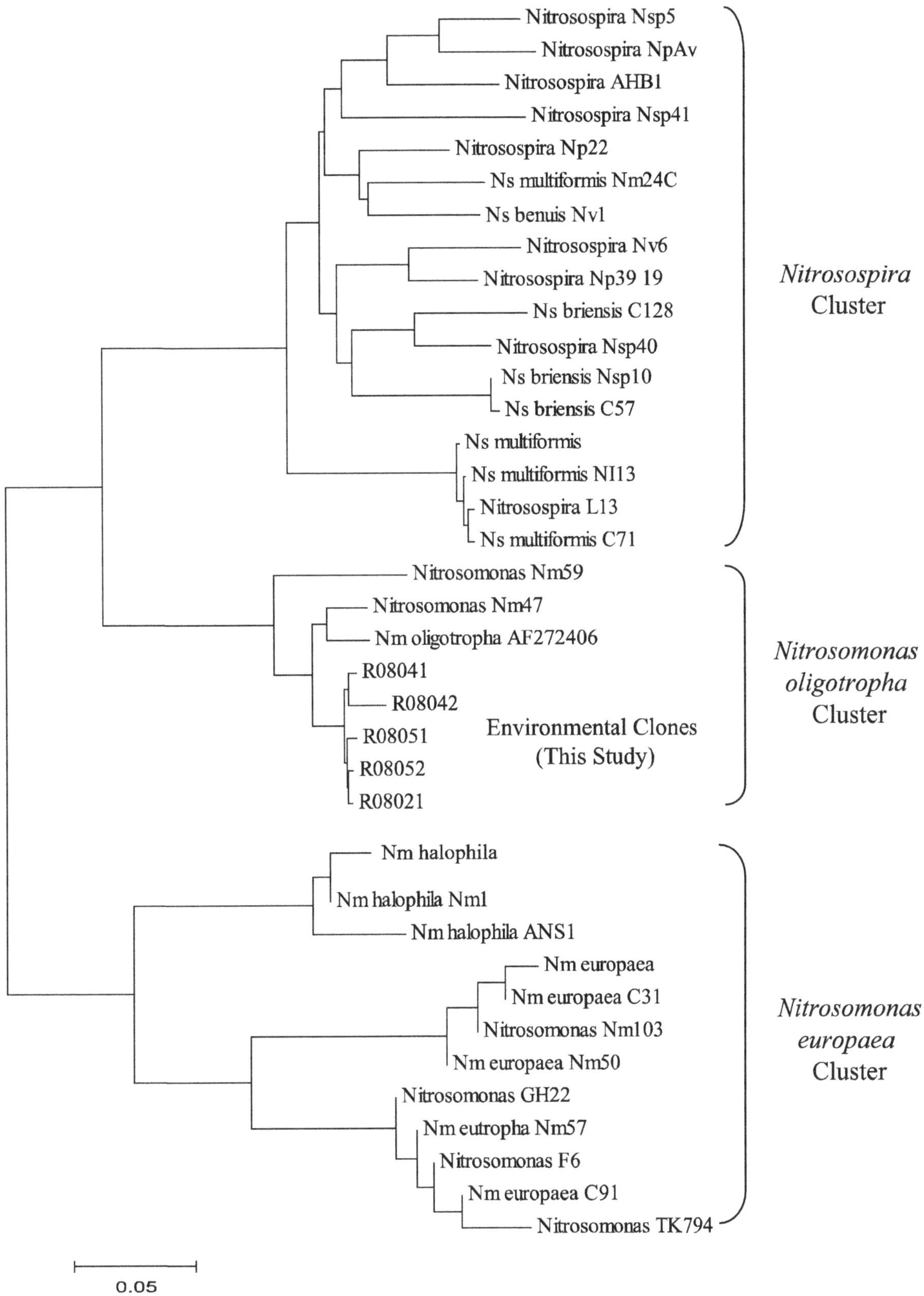

Figure 5.6. Neighbor-joining tree based on *amoA* sequences from β-subgroup AOB.

Standard curves for the three target AOB strains were generated with 10-fold serial dilutions of plasmids containing the respective *amoA* gene fragment (Figures 5.7, 5.8, and 5.9 for *Nm. europaea*, *Ns. multiformis*, and *Nm. oligotropha*-like AOB). The most concentrated plasmid DNA preparations used in the standard curves were 1.37×10^8, 1.52×10^8 and 1.05×10^8 copies for *Nm. europaea*, *Ns. multiformis*, and *Nm. oligotropha*, respectively. Template DNA concentrations greater than 3.5×10^8 copies were found to inhibit amplification, observed in increasing Ct values relative to lower concentrations. In all cases, linear calibration curves were obtained with R^2 values up to 0.98 and linear ranges of seven orders of magnitude. As shown in Figures 5.7, 5.8, and 5.9, the slopes of the standard curves were -3.6393, -3.6946, and -3.9711, which corresponds to amplification efficiencies of 88%, 86%, and 85%, respectively. The detection limits of the PCR assays were 137, 152, and 105 gene copies per PCR reaction for *Nm. europaea*, *Ns. multiformis*, and *Nm. oligotropha*, respectively. Similar amplification and detection characteristics have been reported by previous studies using real-time PCR. Kolb *et al.* (2003) obtained a detection level of approximately 100 *pmoA* gene copies per reaction, and Okano *et al.* (2004) showed a detection limit of 98.3 *amoA* gene copies per PCR reaction.

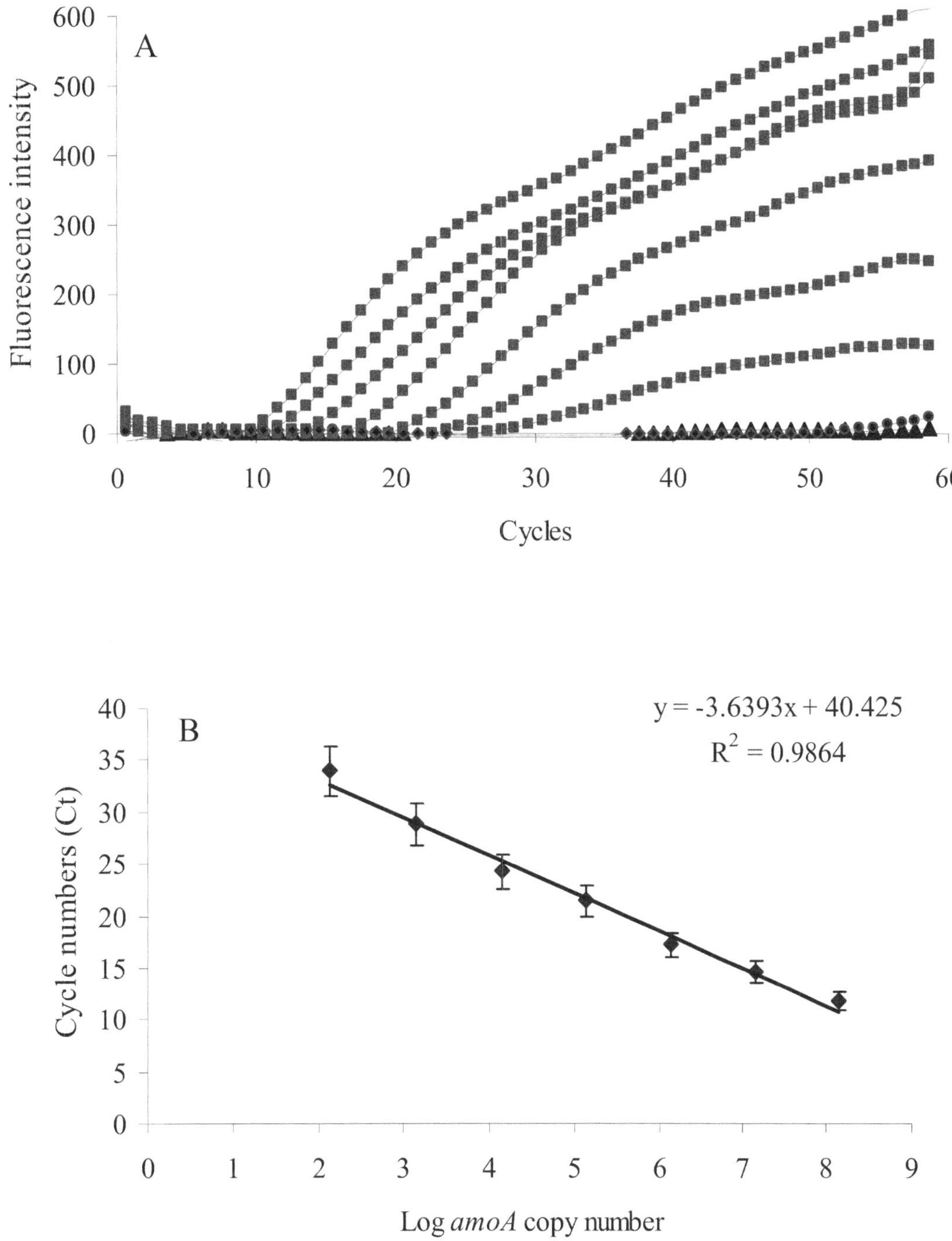

Figure 5.7 The amplification plot (A) and corresponding standard curve (B) for *amoA*-targeted real-time PCR of *Nm. europaea*. For panel (A): ■, *Nm. europaea*; ▲, *Ns. multiformis*; ●, *Nm. oligotropha*-like AOB; ♦, no DNA.

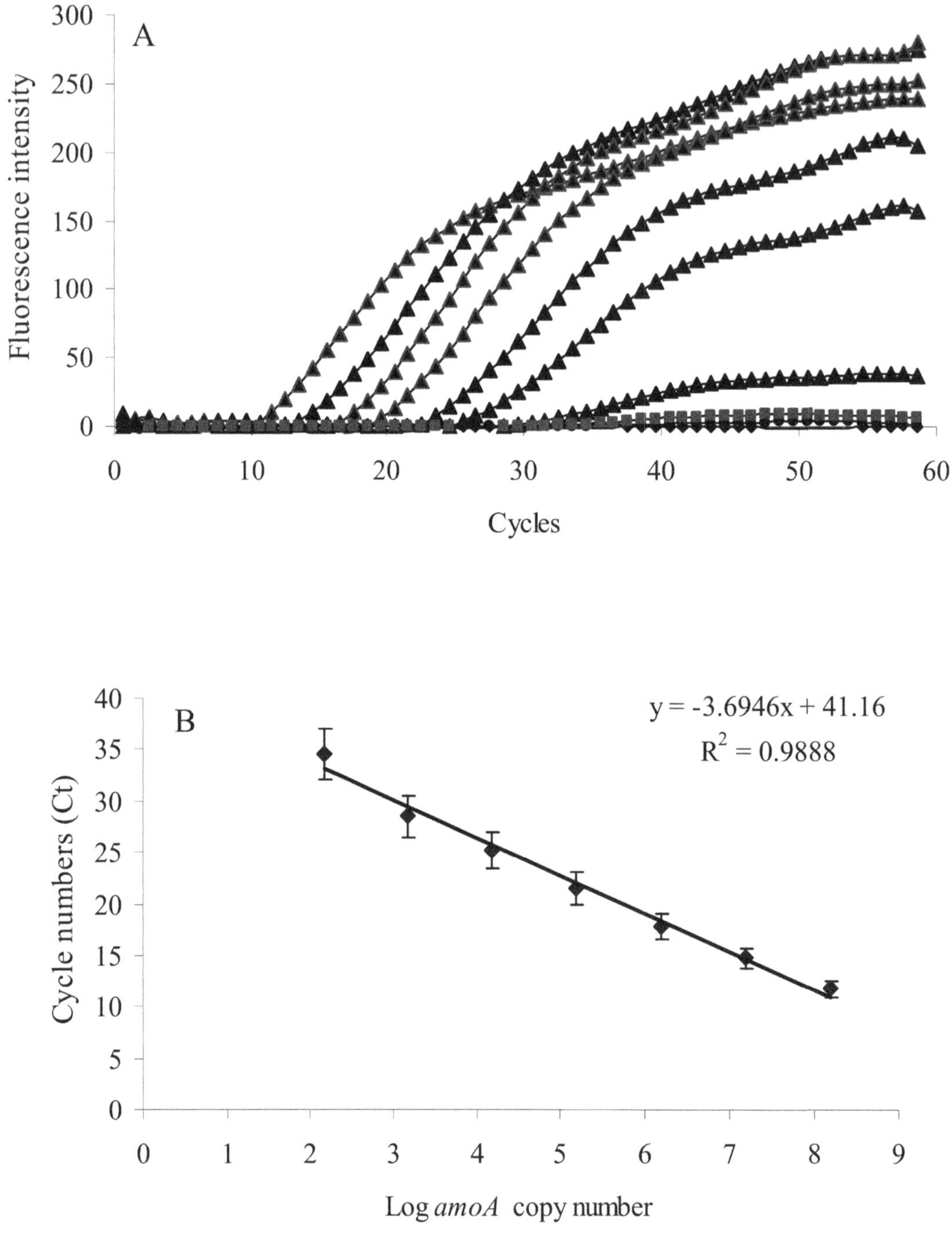

Figure 5.8 The amplification plot (A) and corresponding standard curves (B) for *amoA*-targeted real-time PCR of *Ns. multiformis*. For panel (A): ▲, *Ns. multiformis* ; ■, *Nm. europaea*; ●, *Nm. oligotropha*-like AOB; ♦, no DNA.

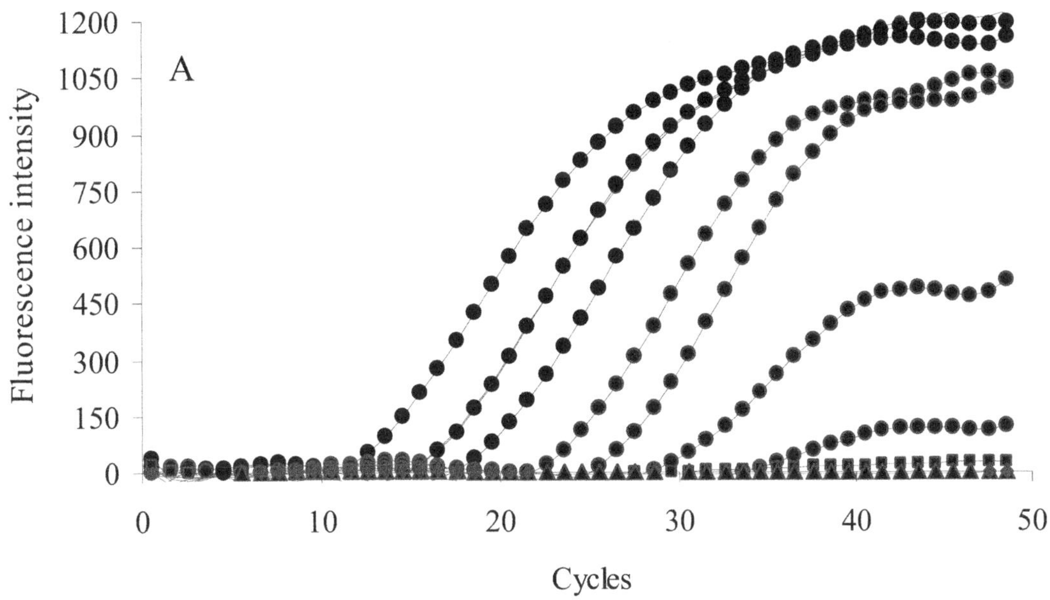

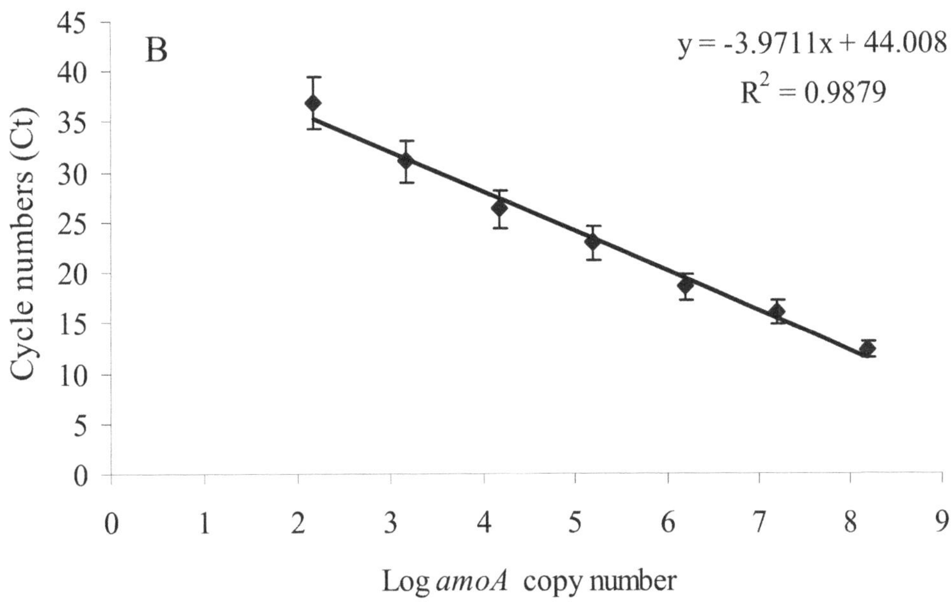

Figure 5.9 The amplification plot (A) and corresponding standard curves (B) for *amoA*-targeted real-time PCR of *Nm. oligotropha*-like AOB. For panel (A): ●, *Nm. oligotropha*-like AOB; ■, *Nm. europaea*; ▲, *Ns. multiformis*; ♦, no DNA.

MONITORING OF FULL-SCALE SYSTEMS

As noted in Chapter 3, samples were collected in 2004 from San Francisco Public Utilities Commission (SFPUC), Ann Arbor Water Utilities (AAWU), Pinellas County Utilities (PCU), Alameda County Water District (ACWD), and Louisville Water Company (LWC). In 2005, samples were collected from SFPUC and the Washington DC Water and Sewer Authority (DCWASA). The results from each sampling location are presented in detail below. With the exception of the ACWD samples and one site in the single LWC sampling event, AOB were generally below the detection level in the other systems during the first year of sampling. The lack of AOB proliferation at these other sites was consistent with the chemical monitoring data, which did not show the typical symptoms of nitrification. During the second year of monitoring, both systems showed chemical evidence of nitrification as expected, but AOB were not detected in the SFPUC system until after the water chemistry had changed. All AOB detections throughout this project were with the *Nm. oligotropha* probe; *Nitrosospira* and non-*oligotropha Nitrosomonas* species were not detected in any sample. This exclusive finding of *Nm. oligotropha* is consistent with previous studies of AOB diversity in chloraminated distribution systems (Regan, Harrington, and Noguera 2002; Regan et al. 2003). For AOB enumeration by real-time PCR, the presence of PCR inhibitors in the environmental samples was tested by adding cloned *amoA* fragments from *Nm. europaea*, *Nitrosospira*, and *Nm. oligotropha*-like AOB to aliquots of each sample. The extent of inhibition of spiked samples was below 5%.

San Francisco Public Utilities Commission Results

The SFPUC system was sampled at two locations, Sunset North (Table 5.1) and Stanford Heights (Table 5.2), beginning just two weeks after the system conversion to chloramines on February 2, 2004. Sunset South (Table 5.3) was added to the sampling locations in June 2004. The chloramine residual at these locations averaged 1.96 ± 0.23 mg/l and the average pH was 8.36 ± 0.45. AOB were detected only with the *Nm. oligotropha* real-time PCR assay at Sunset North and Sunset South between June and August. These detections were at levels just below the quantification limit, but given the nondetectable AOB background levels, they may indicate some degree of AOB growth. The June 16 detection in the Sunset North reservoir was accompanied by a slight decrease in residual and a jump in HPC, and sampling a week later showed an increase in nitrite to 50 µg/L. The June 23 detection of AOB at Sunset South (when sampling commenced at this location) was accompanied by an elevated nitrite concentration, and the trace AOB detections in August corresponded to a gradual and slight decline in total chlorine concentration. The system's operational conditions (including control strategies to limit free ammonia levels) apparently were not conducive to a proliferation of AOB. One factor that may have curtailed AOB growth was the generally low temperatures. Although AOB have been detected in chloraminated systems at temperatures below 10°C (Cunliffe 1991), Wolfe et al. (1990) reported a strong correlation between AOB growth in a chloraminated reservoir and temperature above 18°C.

Table 5.1
Chemical and biological data from Sunset North of SFPUC

Date	pH	Temp (C)	HPC (CFU/ml)	Total Chlorine (mg/L)	Free ammonia (mg/L)	Nitrite (mg/L)	*Nm. oligotropha* (copies/ml)	MPN (index/ml)
2/17/04			127				<4.2	<0.2
3/9/04			<10	2.10	0.03	0.004	<4.2	<0.2
3/24/04	8.03	13.0	10	2.25	0.01	0.008	<4.2	<0.2
4/7/04	8.36	11.7	20	2.25	0.05	0.005	<4.2	<0.2
4/21/04	8.33	11.5	10	2.19	0.07	0.002	<4.2	<0.2
5/5/04	9.10	13.2	<10	2.02	0.04	0.006	<4.2	<0.2
5/19/04	8.69	14.0	10	1.47	0.09	0.002	<4.2	<0.2
6/2/04	8.76	12.3	10	2.12	0.06	0.004	<4.2	<0.2
6/16/04	8.69	13.3	2425	1.89	0.05	0.007	2.6*	<0.2
6/23/04	8.23	12.7	10	2.16	0.01	0.050	2.5*	<0.2
6/30/04	8.83	13.6	90	1.93	0.07	0.005	<4.2	<0.2
7/14/04	8.12	14.3	10	1.85	0.05	0.004	<4.2	<0.2
7/28/04	8.40		10	1.92	0.00	0.002	<4.2	<0.2
8/11/04	8.33	14.3	<10	1.88	0.00	0.001	<4.2	<0.2
8/24/04	8.32	19.9	10	1.96	0.00	0.003	2.4*	<0.2
9/8/04			10				<4.2	<0.2
9/22/04	8.45	16.5	<10	2.12	0.08	0.007	<4.2	<0.2

* Below the linear quantitation limit

Table 5.2
Chemical and biological data from Stanford Heights of SFPUC

Date	pH	Temp (C)	HPC (CFU/ml)	Total Chlorine (mg/L)	Free ammonia (mg/L)	Nitrite (mg/L)	*Nm. oligotropha* (copies/ml)	MPN (index/ml)
2/17/04			43				<4.2	<0.2
3/9/04			10	2.40	0.04	0.004	<4.2	<0.2
3/24/04	7.96	13.3	85	2.29	0.00	0.006	<4.2	<0.2
4/7/04	8.00	16.5	<10	2.21	0.05	0.006	<4.2	<0.2
4/21/04	8.70	13.2	20	2.05	0.08	0.004	<4.2	<0.2
5/5/04	8.76	14.8	10	1.94	0.04	0.007	<4.2	<0.2
5/19/04	8.47	14.3	<10	1.96	0.07	0.000	<4.2	<0.2
6/2/04	8.76	12.5	10	1.98	0.03	0.003	<4.2	<0.2
6/16/04	7.98	18.3	400	1.28	0.13	0.002	<4.2	<0.2
6/30/04	8.73	15.6	10	1.95	0.04	0.005	<4.2	<0.2
7/14/04	8.14	16.2	20	1.90	0.08	0.004	<4.2	<0.2
7/28/04	8.61		20	1.85	0.00	0.004	<4.2	<0.2
8/11/04	8.43	16.2	<10	1.86	0.04	0.001	<4.2	<0.2
8/24/04	8.15	19.6	<10	1.88	0.01	0.004	<4.2	<0.2
9/8/04			<10				<4.2	<0.2
9/22/04	8.45	16.5	30	1.32	0.15	0.001	<4.2	<0.2

Table 5.3
Chemical and biological data from Sunset South of SFPUC

Date	pH	Temp (C)	HPC (CFU/ml)	Total Chlorine (mg/L)	Free ammonia (mg/L)	Nitrite (mg/L)	*Nm. oligotropha* (copies/ml)	MPN (index/ml)
6/23/04	8.54	13.6	20	2.03	0.00	0.070	2.7*	<0.2
6/30/04	8.84	13.6	10	1.94	0.04	0.004	2.0*	<0.2
7/14/04	8.08	14.5	20	2.00	0.10	0.003	<4.2	<0.2
7/28/04	8.36		10	1.88	0.00	0.003	<4.2	<0.2
8/4/04	7.67	14.9	<10	1.92	0.04	0.009	2.7*	<0.2
8/11/04	8.49	14.7	10	1.87	0.01	0.004	3.2*	<0.2
8/24/04	8.33	19.9	<10	1.74	0.03	0.003	1.9*	<0.2
9/9/04			20				<4.2	<0.2
9/22/04	8.35	16.6	20	2.09	0.09	0.007	<4.2	<0.2

* Below the linear quantitation limit

Ann Arbor Water Utilities Results

Samples were collected from four tap locations in the AAWU system (Tables 5.4 - 5.7). The average monochloramine concentration over the sampling period was 2.2 ± 0.3 mg/l, and the average pH was 9.3 ± 0.1. The maintenance of a high pH is part of the nitrification control strategy employed by AAWU (Skadsen 2002). At all four sampling locations, AOB were detected with the MPN assay at very low levels, and *Nm. oligotropha* were detected by real-time PCR at three of the locations at concentrations below the quantification limit. There was no clear evidence of AOB growth in AAWU's distribution system over the four-month period. The elevated nitrate concentrations can be attributed to the raw water quality, and were not principally associated with nitrification. For nitrite, there were low levels present in the raw water and an increase across the filters, with a slight increase observed in the distribution system (correspondence with utility). With this detectable baseline nitrite concentration, it was not possible to specifically correlate nitrite fluctuations with AOB growth.

Pinellas County Utilities Results

Four distribution system locations were included in the PCU sampling (Tables 5.8 – 5.11). This utility switched to chloramines in May 2002 and was added to our sampling program with the expectation that we would encounter periods of nitrification. While the system consistently had a high temperature (28 ± 2 °C) and was expected to nitrify, the oxidized nitrogen and AOB monitoring data did not show evidence of nitrification. AOB were only detected on two sampling events at two locations, and the concentration was below the quantitation limit. There were no MPN detections of AOB. A possible explanation for the absence of AOB proliferation in this system is the high chloramine residual of 2.7 ± 0.6 mg/l, and the practice of distribution system flushing. The maintenance of the system also included periodic breakpoint chlorination, including one episode that commenced on October 25, 2004 and is evident in the monochloramine data in Tables 5.10 and 5.11. In addition, the Gulf Beach tank was drained and modified during August and September to reduce the water age in this above-ground storage tank.

Table 5.4
Chemical and biological data from #4 of AAWU

Date	pH	Temp (C)	HPC (CFU/ml)	Monochloramine (mg/L)	Free ammonia (mg/L)	Nitrite (mg/L)	Nitrate (mg/L)	*Nm. oligotropha* (copies/ml)	MPN (index/ml)
6/29/04	9.3	20.3	656	1.9	0.17	0.026	0.78	<4.2	0.2
7/15/04	9.3	23.6	232	2.2	0.18	0.04	0.41	<4.2	0.4
8/4/04	9.2	23.0	2793	1.6	0.13	0.057	0.4	2.2*	2.2
8/17/04	9.3	19.2	656	2.4	0.19	0.066	0.4	2.1*	5
9/1/04	9.4	20.6	529	2.5	0.12	0.054	0.35	<4.2	0.4
9/22/04			1025	2.6	0.18	0.037	0.38	<4.2	0.2
10/13/04			855	2.5	0.18	0.037	0.33	<4.2	<0.2
11/1/04			126	2.6	0.21	0.048	0.35	<4.2	<0.2

* Below the linear quantitation limit

Table 5.5
Chemical and biological data for #5 Cranbrook of AAWU

Date	pH	Temp (C)	HPC (CFU/ml)	Monochloramine (mg/L)	Free ammonia (mg/L)	Nitrite (mg/L)	Nitrate (mg/L)	*Nm. oligotropha* (copies/ml)	MPN (index/ml)
6/29/04	9.3	20.3	1992	2.0	0.19	0.038	1.21	<4.2	0.4
7/15/04	9.3	23.6	1971	2.3	0.18	0.038	0.39	<4.2	0.4
8/4/04	9.2	23.0	1368	2.1	0.13	0.034	0.41	<4.2	0.4
8/17/04	9.3	19.2	770	2.2	0.19	0.036	0.41	<4.2	0.4
9/1/04	9.4	20.6	512	2.3	0.13	0.027	0.35	<4.2	0.2
9/22/04			1254	2.7	0.20	0.027	0.36	<4.2	0.4
10/13/04			2964	2.5	0.18	0.027	0.34	<4.2	<0.2
11/1/04			513	2.9	0.19	0.042	0.36	<4.2	<0.2

Table 5.6
Chemical and biological data from #8 City Hall of AAWU

Date	pH	Temp (C)	HPC (CFU/ml)	Monochloramine (mg/L)	Free ammonia (mg/L)	Nitrite (mg/L)	Nitrate (mg/L)	*Nm. oligotropha* (copies/ml)	MPN (index/ml)
6/29/04	9.3	20.3	7	2.6	0.06	0.027	0.66	<4.2	0.4
7/15/04	9.3	23.6	12	2.1	0.25	0.035	0.44	2.2*	1.3
8/4/04	9.2	23.0	8	2.7	0.11	0.025	0.38	2.1*	1.3
8/17/04	9.3	19.2	6	2.9	0.12	0.023	0.38	<4.2	0.2
9/1/04	9.4	20.6	4	2.0	0.09	0.016	0.38	<4.2	0.2
9/22/04			11	3.0	0.07	0.015	0.35	3.4*	1.3
10/13/04			16	3.0	0.14	0.021	0.33	<4.2	<0.2
11/1/04			513	2.9	0.19	0.042	0.36	<4.2	<0.2

* Below the linear quantitation limit

Table 5.7
Chemical and biological data from #10 Art of AAWU

Date	pH	Temp (C)	HPC (CFU/ml)	Monochloramine (mg/L)	Free ammonia (mg/L)	Nitrite (mg/L)	Nitrate (mg/L)	*Nm. oligotropha* (copies/ml)	MPN (index/ml)
6/29/04	9.3	20.3	132	1.8	0.14	0.11	0.75	<4.2	5
7/15/04	9.3	23.6	92	2.2	0.17	0.09	0.45	2.4*	5
8/4/04	9.2	23.0	192	1.7	0.18	0.09	0.35	2.5*	5
8/17/04	9.3	19.2	58	2.2	0.13	0.10	0.35	<4.2	0.3
9/1/04	9.4	20.6	9	2.0	0.14	0.02	0.36	<4.2	0.4
9/22/04			51	2.7	0.12	0.021	0.35	4.5	3
10/13/04			6	2.7	0.18	0.027	0.33	<4.2	0.2
11/1/04			45	2.8	0.26	0.032	0.33	<4.2	<0.2

Table 5.8

Chemical and biological data for Gulf Beach Tank Effluent of PCU

Date	pH	Temp. (C)	HPC (CFU/ml)	Total Chlorine (mg/L)	Monochloramine (mg/L)	Free ammonia (mg/L)	Nitrite (mg/L)	Nitrate (mg/L)	*Nm. oligotropha* (copies/ml)	MPN (index/ml)
7/22/04	7.73	30.1	10	3.1	2.7	0.38	0.019	0.03	<4.2	<0.2
7/27/04	7.82	28.9	1	3.1	3.0	0.55	0.120	0.03	<4.2	<0.2
8/3/04	7.83	28.2	1	3.0	2.8	NA	0.007	0.02	<4.2	<0.2
8/10/04	7.77	28.6	1	3.3	2.9	0.38	0.012	0.01	<4.2	<0.2
8/17/04	7.83	28.9	<1	2.7	4.0	0.37	0.008	0.02	<4.2	<0.2
8/24/04*										
9/23/04	7.90	29.4	<1	2.2	1.7	0.37	0.009	0.02	<4.2	<0.2
9/30/04	7.87	29.9	<1	2.7	2.0	NA	0.013	0.03	<4.2	<0.2
10/5/04	7.75	28.5	<1	3.2	2.1	0.27	0.006	0.02	<4.2	<0.2
10/12/04	7.68	26.9	<1	2.9	2.1	0.34	0.018	0.03	<4.2	<0.2
10/19/04	7.76	26.5	<1	2.6	1.6	0.46	0.009	0.04	<4.2	<0.2
10/26/04	7.81	28.7	<1	3.0	2.1	0.36	0.023	0.07	<4.2	<0.2
11/2/04	7.50	28.1	290†	3.6	0.2	0.11	0.012	0.11	<4.2	<0.2

* Tank drained and modified to reduce water age

† Standard Method 9215 D (Membrane filter method)

Table 5.9

Chemical and biological data for #225 of PCU

Date	pH	Temp. (C)	HPC (CFU/ml)	Total Chlorine (mg/L)	Monochloramine (mg/L)	Free ammonia (mg/L)	Nitrite (mg/L)	Nitrate (mg/L)	*Nm. oligotropha* (copies/ml)	MPN (index/ml)
7/22/04	7.84	30.5	2	3.1	3.1	0.02	0.018	0.04	<4.2	<0.2
7/27/04	7.96	30.0	4	3.1	2.9	0.53	0.130	0.03	<4.2	<0.2
8/3/04	7.85	29.5	<1	3.2	2.7	NA	0.017	0.02	<4.2	<0.2
8/10/04	7.83	29.9	1	3.0	2.8	0.33	0.011	0.02	<4.2	<0.2
8/17/04	7.86	30.0	<1	3.0	3.1	0.29	0.009	0.02	<4.2	<0.2
8/24/04	7.86	30.2	<1	2.7	2.6	0.41	0.019	0.02	<4.2	<0.2
8/31/04	7.99	30.6	8	3.2	2.9	0.31	0.011	0.02	2.2*	<0.2
9/8/04	7.87	30.0	<1	2.5	2.0	0.31	0.015	0.03	2.1*	<0.2
9/16/04	7.93	29.3	<1	2.6	2.0	0.24	0.01	0.02	<4.2	<0.2
9/23/04	7.83	28.8	<1	2.7	2.3	0.46	0.011	0.02	<4.2	<0.2
9/30/04	7.80	28.9	2	3.1	2.3	0.36	0.009	0.02	<4.2	<0.2
10/5/04	7.75	28.6	<1	2.8	1.9	0.31	0.01	0.03	<4.2	<0.2
10/12/04	7.65	28.7	1	3.2	2.2	0.38	0.006	0.02	<4.2	<0.2
10/19/04	7.75	27.3	1	2.8	1.9	0.38	0.032	0.03	<4.2	<0.2
10/26/04	7.81	29.1	<1	3.3	2.3	0.39	0.024	0.04	<4.2	<0.2
11/2/04	7.48	28.1	1 †	3.8	0.3	<0.02	0.018	0.04	<4.2	<0.2

* Below the linear quantitation limit

† Standard Method 9215 D (Membrane filter method)

Table 5.10
Chemical and biological data for 60" Intertie of PCU

Date	pH	Temp (C)	HPC (CFU/ml)	Total Chlorine (mg/L)	Monochloramine (mg/L)	Free ammonia (mg/L)	Nitrite (mg/L)	Nitrate (mg/L)	*Nm. oligotropha* (copies/ml)	MPN (index/ml)
7/22/04	7.76	26.5	1	4.4	4.0	0.24	0.017	0.01	<4.2	<0.2
7/27/04	7.80	26.5	17	4.0	3.0	0.28	<0.005	0.01	<4.2	<0.2
8/3/04	7.72	27.8	<1	4.7	4.5	NA	0.005	0.01	<4.2	<0.2
8/10/04	7.73	26.0	<1	4.1	3.1	0.26	<0.005	<0.01	<4.2	<0.2
8/17/04	7.86	26.7	<1	4.2	3.3	0.19	0.005	0.02	<4.2	<0.2
8/24/04	7.76	27.3	<1	4.1	3.1	<0.02	<0.005	0.29	<4.2	<0.2
8/31/04	7.89	34.0	<1	4.2	3.0	0.17	<0.005	0.01	<4.2	<0.2
9/8/04	7.69	27.2	<1	4.6	3.2	0.24	0.007	0.01	<4.2	<0.2
9/16/04	7.60	28.4	<1	4.1	2.9	0.14	0.006	0.01	<4.2	<0.2
9/23/04	7.92	26.8	<1	4.8	3.3	0.24	0.007	0.03	<4.2	<0.2
9/30/04	7.94	27.0	<1	4.2	3.0	0.17	0.007	0.03	<4.2	<0.2
10/5/04	7.82	27.8	<1	4.0	3.2	0.19	<0.005	0.01	<4.2	<0.2
10/12/04	7.81	23.5	<1	4.3	2.7	0.20	0.011	0.06	<4.2	<0.2
10/19/04	7.85	24.2	<1	4.3	3.2	0.11	0.030	0.04	<4.2	<0.2
10/26/04	7.67	30.9	<1	4.9	0.2	0.05	0.032	0.06	<4.2	<0.2
11/2/04	7.71	27.2	2 *	5.5	0.3	0.03	0.023	0.05	<4.2	<0.2

* Standard Method 9215 D (Membrane filter method)

Table 5.11
Chemical and biological data for Shoreline Drive of PCU

Date	pH	Temp. (C)	HPC (CFU/ml)	Total Chlorine (mg/L)	Monochloramine (mg/L)	Free ammonia (mg/L)	Nitrite (mg/L)	Nitrate (mg/L)	*Nm. oligotropha* (copies/ml)	MPN (index/ml)
7/22/04	7.75	27.9	<1	3.7	3.5	0.35	0.019	0.02	<4.2	<0.2
7/27/04	7.78	29.3	<1	3.3	2.4	<0.02	0.006	0.01	<4.2	<0.2
8/3/04	7.70	29.3	<1	3.8	3.6	NA	0.005	0.01	<4.2	<0.2
8/10/04	7.82	27.2	<1	3.4	2.6	0.28	0.007	<0.01	<4.2	<0.2
8/17/04	7.84	27.3	<1	3.6	2.6	0.32	0.011	0.02	<4.2	<0.2
8/24/04	7.72	29.0	<1	4.1	2.7	<0.02	0.009	0.30	2.5*	<0.2
8/31/04	7.79	34.3	2	3.3	2.5	0.29	0.008	0.02	2.7*	<0.2
9/8/04	7.71	27.6	<1	3.1	2.3	0.31	0.007	0.02	<4.2	<0.2
9/16/04	7.56	27.9	<1	3.5	2.5	0.24	<0.005	0.06	<4.2	<0.2
9/23/04	7.71	26.9	<1	3.5	2.6	0.28	0.007	0.01	<4.2	<0.2
9/30/04	7.80	26.8	<1	3.7	2.7	0.21	0.015	0.01	<4.2	<0.2
10/5/04	7.78	27.7	<1 †	3.5	2.7	0.21	0.008	0.01	<4.2	<0.2
10/12/04	7.79	25.8	4	3.9	2.7	0.29	0.019	0.04	<4.2	<0.2
10/19/04	7.76	25.7	<1	3.8	2.4	NA	0.023	0.06	<4.2	<0.2
10/26/04	7.81	28.2	<1	1.2	0.8	0.11	0.009	0.04	<4.2	<0.2
11/2/04	7.63	27.4	450 ‡	2.8	0.3	0.07	0.015	0.02	<4.2	<0.2

* Below the linear quantitation limit
† Beyond hold time
‡ Standard Method 9215 D (Membrane filter method)

Alameda County Water District Results

In the ACWD system, samples were collected from their 15 MG finished water storage reservoir located in Fremont, California. Samples were collected at reservoir ports 2, 4, and 5, as designated in Figure 3.1, and the data are presented in Tables 5.12 to 5.14. The Alameda Reservoir has historically experienced a reduction in disinfectant residual as a result of nitrification, even during the cold temperature season. Our monitoring of this reservoir spanned several nitrification episodes, along with ACWD's response of breakpoint chlorination on June 15, July 12, July 27, July 29, and August 31, 2004. The District also explored the implementation of a ClorTec Reservoir Management System™ (RMS), which adds and mixes chlorine and ammonia to boost the chloramine residual. This booster chloramination system began on September 27, 2004, was turned off on October 16, 2004 and resumed on November 11, 2004.

The chemical and biological data collected from ACWD during this study showed clear evidence of repeated nitrification problems. This was apparent in the traditional indicators of a reduction in chloramine residual and an increase in nitrite, as shown in Figure 5.10, and the growth of AOB, presented in Figure 5.11. The system response to breakpoint chlorination on

Table 5.12
Chemical and biological data for Reservoir 8 - Location 2 of ACWD

Date	pH	Temp. (C)	HPC (CFU/ml)	Monochloramine (mg Cl2/L)	Free ammonia (mg N/L)	Total ammonia (mg N/L)	Nitrite (mg N/L)	TOC (mg/L)	*Nm. oligotropha* (copies/ml)	MPN (index/ml)
7/6/04	7.86	21.9	1	1.84	0.06	0.42	0.069	1.4	<4.2	0.4
7/13/04	8.44		1	0.80	0.07	0.17	0.037	1.53	29.8	1.7
7/20/04	8.34	18.7	2	0.52	0.04	0.10	0.041	1.5	120.0	3.0
7/27/04	8.36	20.0	2	0.10	0.01	0.02	0.171	1.4	151.3	5.0
8/2/04	8.63	20.7	1	1.07	0.09	0.23	0.008	1.24	<4.2	0.4
8/10/04	8.53	20.8	1	1.70	0.10	0.46	0.012	1.53	<4.2	0.8
8/17/04	8.52	21.2	1	1.75	0.16	0.37	0.160	2.03	7.4	2.3
8/24/04	8.30	22.2	1	0.20	0.01	0.05	0.170	1.75	84.7	24
8/31/04	8.46	21.2	2	0.35	0.03	0.07	0.094	1.64	142.8	22
9/7/04	8.50	23.3	1	1.01	0.12	0.29	0.060	1.7	303.4	30
9/14/04	8.40	19.6	<1	0.55	0.07	0.08	0.153	2.2	511.3	50
9/22/04			<1	0.30	0.05		0.039	2.2	429.7	24
9/29/04	8.27	17.6	<1	1.12	0.00	0.22	0.009	2.2	<4.2	0.4
10/6/04	8.35	18.4	<1	1.93	0.15	0.42	0.026		<4.2	0.2
10/12/04	8.43	18.5	<1	2.50	0.08	0.41	0.006		<4.2	<0.2
10/19/04	8.46	17.4	<1	2.27	0.17	0.45	0.009		<4.2	<0.2
10/26/04	8.47	15.5	<1	1.57	0.16	0.45	0.008	1.9	<4.2	<0.2
11/2/04	8.39	16.8	<1	1.43	0.27	0.45	0.006	2	<4.2	<0.2
11/9/04	8.38	15.7	<1	1.30	0.22	0.34	0.029	2	56.4	2.3
11/16/04	8.46	17.0	<1	2.38	0.19	0.60	0.012	2.1	8.8	<0.2
11/23/04	8.54	14.5	<1	2.39	0.15	0.47	0.011		<4.2	<0.2
11/30/04	8.50			2.41	0.14	0.43	0.009		<4.2	<0.2
12/7/04	8.50			2.26	0.15		0.008		<4.2	<0.2
12/14/04				2.32	0.08		0.009		<4.2	<0.2
12/21/04				1.72	0.03	0.45	0.005		<4.2	<0.2

Table 5.13
Chemical and biological data for Reservoir 8 - Location 4 of ACWD

Date	pH	Temp. (C)	HPC (CFU/ml)	Monochloramine (mg Cl2/L)	Free ammonia (mg N/L)	Total ammonia (mg N/L)	Nitrite (mg N/L)	TOC (mg/L)	*Nm. oligotropha* (copies/ml)	MPN (index/ml)
6/2/04				0.62	0.58	0.24	0.052		<4.2	0.4
6/9/04	8.50	20.0	1	0.40	0.04	0.12	0.045	1.8	89.8	30
6/11/04	8.72	21.7	1	0.35	0.04	0.08	0.023	1.8	44.7	17
6/15/04	8.79	20.2	1	0.67	0.06	0.13	0.071	1.6	<4.2	13
6/22/04	8.82	20.2	1	1.35	0.10	0.33	0.010	1.43	<4.2	0.4
6/29/04	8.90	19.8	1	1.76	0.14	0.41	0.021	1.37	142.8	13
7/6/04	8.76	19.5	1	1.73	0.12	0.44	0.056	1.4	95.2	22
7/13/04	8.47		1	0.87	0.06	0.19	0.046	1.53	142.8	50
7/20/04	8.34	18.7	2	0.49	0.03	0.09	0.042	1.4	286.3	13
7/27/04	8.44	20.3	7	0.12	0.05	0.03	0.172	1.3	134.7	8
8/2/04	8.50	21.1	1	1.06	0.10	0.27	0.008	1.21	<4.2	0.4
8/10/04	8.61	20.8	1	1.77	0.09	0.44	0.012	1.57	<4.2	0.8
8/17/04	8.63	20.9	1	1.71	0.12	0.40	0.140	1.94	17.7	5
8/24/04	8.45	22.4	1	0.21	0.03	0.05	0.167	2.13	361.1	22
8/31/04	8.49	21.1	1	0.36	0.05	0.06	0.095	2.74	482.5	22
9/7/04	8.63	23.5	1	0.96	0.12	0.28	0.064	1.7	405.5	35
9/14/04	8.31	19.9	<1	0.52	0.00	0.09	0.160	2.5	574.2	50
9/22/04			1	0.33	0.04		0.047	2.2	482.5	30
9/29/04	8.28	16.9	<1	0.77	0.00	0.18	0.007	2.2	<4.2	0.4
10/6/04	8.40	16.9	<1	1.82	0.13	0.37	0.019		<4.2	0.2
10/12/04	8.41	18.4	<1	2.56	0.06	0.39	0.006		<4.2	<0.2
10/19/04	8.46	16.9	3	2.24	0.15	0.46	0.006		<4.2	<0.2
10/26/04	8.51	15.5	<1	1.58	0.19	0.43	0.006	2.1	<4.2	<0.2
11/2/04	8.36	15.7	<1	1.19	0.28	0.47	0.008	2.1	<4.2	<0.2
11/9/04	8.36	15.7	4	1.25	0.19	0.39	0.033	2.1	39.8	2.7
11/16/04	8.48	15.9	<1	2.24	0.22	0.59	0.024	2	14.0	<0.2
11/23/04	8.54	14.3	<1	2.32	0.18	0.48	0.010		<4.2	<0.2
11/30/04	8.50			2.49	0.10	0.51	0.007		<4.2	<0.2
12/7/04	8.50			2.28	0.14		0.004		<4.2	<0.2
12/14/04				2.47	0.04		0.007		<4.2	<0.2
12/21/04				2.36	0.04	0.45	0.005		<4.2	<0.2

July 27-29 and August 31, as well as the effect of booster chloramination beginning onSeptember 27, are evident in the increase in total chlorine residual concentration and limit in nitrite production. The data from this reservoir show approximately a one-month period of effectiveness for breakpoint chlorination. The effects of nitrification also appear in the nitrite and real-time PCR data approximately one month after the booster chloramination system was turned off in mid-October. The chloramine residual also dropped during this time. One of the goals of this study was to evaluate whether the rapid molecular quantification of AOB could serve as an early indication of the development of a nitrification problem. A comparison of the real-time PCR data with nitrite and monochloramine are shown in Figures 5.11 and 5.12, respectively, for one of the ACWD reservoir ports. Inspection of Figures 5.11 and 5.12 shows that in early July, the AOB concentration increased and the chloramine residual decreased, while

Table 5.14
Chemical and biological data for Reservoir 8 - Location 5 of ACWD

Date	pH	Temp. (C)	HPC (CFU/ml)	Monochloramine (mg Cl2/L)	Free ammonia (mg N/L)	Total ammonia (mg N/L)	Nitrite (mg N/L)	TOC (mg/L)	*Nm. oligotropha* (copies/ml)	MPN (index/ml)
6/9/04	8.63	20.1	1	0.39	0.06	0.11	0.053	1.8	59.8	1.7
6/11/04	8.40	21.6	1	0.31	0.04	0.11	0.024	1.8	35.5	1.3
6/15/04	8.65	20.8	1	0.76	0.04	0.14	0.071	1.6	<4.2	0.8
6/22/04	8.71	20.7	1	1.34	0.05	0.29	0.008	1.46	<4.2	0.2
6/29/04	8.77	20.2	1	1.71	0.12	0.41	0.021	1.33	48.8	0.2
7/6/04	8.77	19.3	1	1.79	0.11	0.42	0.049	1.5	37.6	1.3
7/13/04	8.45		1	0.89	0.04	0.17	0.042	1.61	100.8	5.0
7/20/04	8.34	18.7	24	0.49	0.03	0.10	0.036	1.3	180.1	17.0
7/27/04	8.45	20.4	6	0.15	0.00	0.02	0.170	1.3	240.6	8.0
8/2/04	8.55	20.7	4	1.15	0.05	0.23	0.006	1.21	<4.2	0.4
8/10/04	8.57	20.8	1	1.74	0.18	0.44	0.012	1.58	<4.2	0.8
8/17/04	8.48	20.8	1	1.76	0.12	0.39	0.170	1.93	9.9	8.0
8/24/04	8.33	22.3	3	0.22	0.00	0.05	0.166	1.76	100.8	24.0
8/31/04	8.48	21.1	1	0.36	0.04	0.07	0.093	1.67	321.5	50.0
9/7/04	8.56	23.2	1	1.02	0.07	0.28	0.064	1.7	361.1	50.0
9/14/04	8.34	19.9	3	0.37	0.02	0.09	0.157	2.1	644.8	50.0
9/22/04			2	0.33	0.05		0.056	2.1	482.5	30.0
9/29/04	8.24	16.8	<1	0.81	0.04	0.16	0.005	2.2	<4.2	0.4
10/6/04	8.49	17.1	<1	1.95	0.10	0.40	0.022		<4.2	0.2
10/12/04	8.40	18.1	<1	2.53	0.06	0.37	0.005		<4.2	<0.2
10/19/04	8.46	16.8	43	2.30	0.13	0.44	0.007		<4.2	<0.2
10/26/04	8.47	15.7	<1	1.56	0.22	0.41	0.012	1.9	<4.2	<0.2
11/2/04	8.40	15.9	<1	1.34	0.12	0.46	0.010	2	2.3*	<0.2
11/9/04	8.37	15.8	<1	1.44	0.19	0.31	0.033	2	75.4	3.0
11/16/04	8.46	17.6	<1	2.16	0.25	0.57	0.022	2	<4.2	<0.2
11/23/04	8.54	13.8	<1	2.34	0.16	0.49	0.013		<4.2	<0.2
11/30/04	8.50			2.47	0.11	0.48	0.009		<4.2	<0.2
12/7/04	8.50			2.25	0.13		0.003		<4.2	<0.2
12/14/04				2.50	0.09		0.011		<4.2	<0.2
12/21/04				2.35	0.02	0.36	0.004		<4.2	<0.2

* Below the linear quantitation limit

the nitrite concentration showed a two-week lag in a detectable increase. Similarly, in September the AOB population was increasing, while the nitrite concentration concurrently showed a marked decrease. This second incident also coincided with an increasing chloramine residual for two weeks, after which the residual began to decline. The limited results from these two periods support the hypothesis that an increase in AOB concentration can be more reliable than chemical indicators of a pending nitrification problem.

There are two sampling dates that appear to contradict this assertion. On August 17, the nitrite concentration in the reservoir showed a dramatic increase, but the AOB concentration showed only an incremental increase and the chloramine residual remained constant. While the AOB concentration did not show the dramatic change that was evident in the nitrite concentration, it increased from below the detection limit to quantifiable numbers. There is

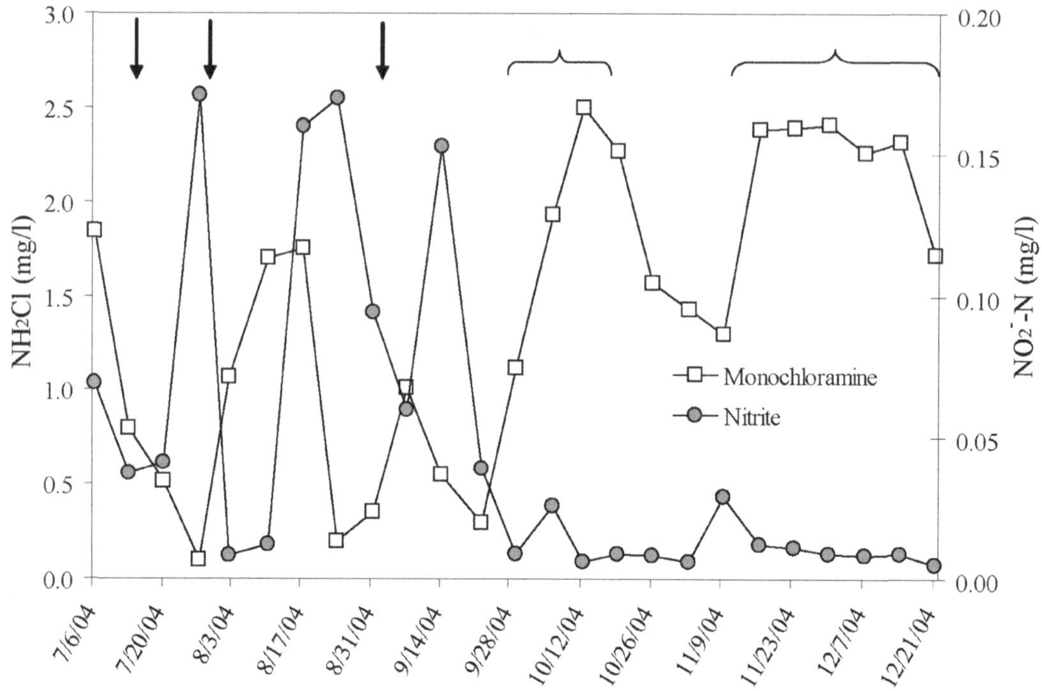

Figure 5.10 Monochloramine and nitrite data from Port 2 of ACWD Reservoir 8. Arrows denote breakpoint chlorination events; brackets denote booster chloramination periods.

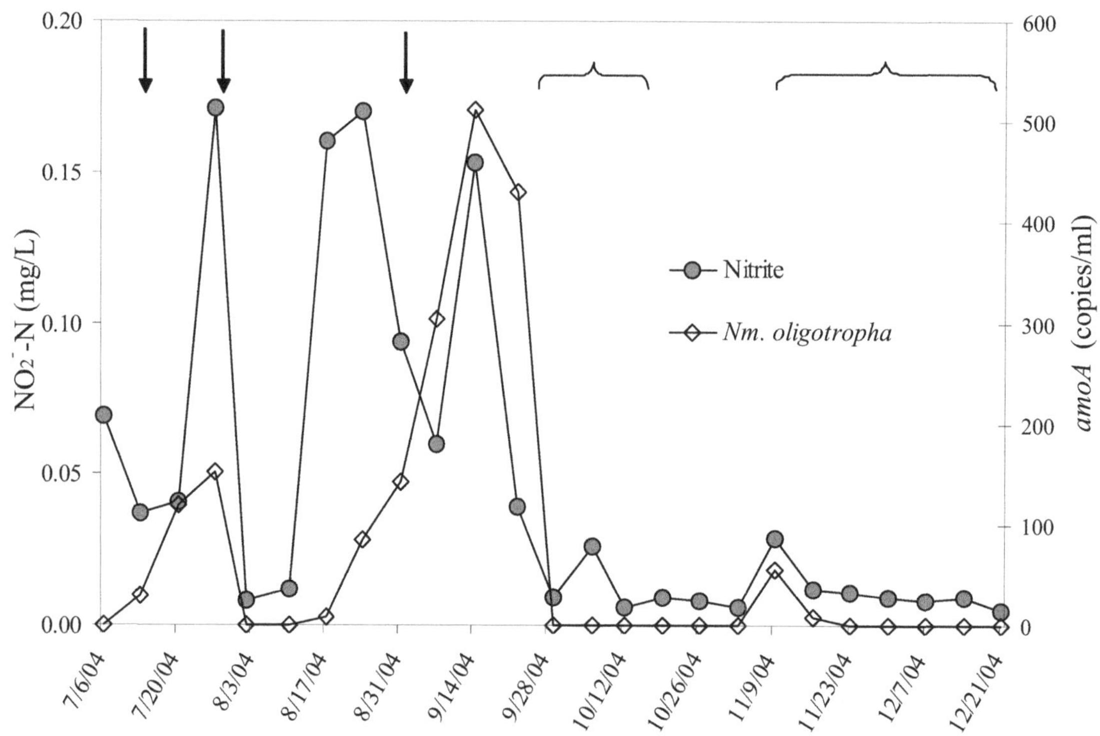

Figure 5.11 Nitrite and real-time PCR data from Port 2 of ACWD Reservoir 8. Arrows denote breakpoint chlorination events; brackets denote booster chloramination periods.

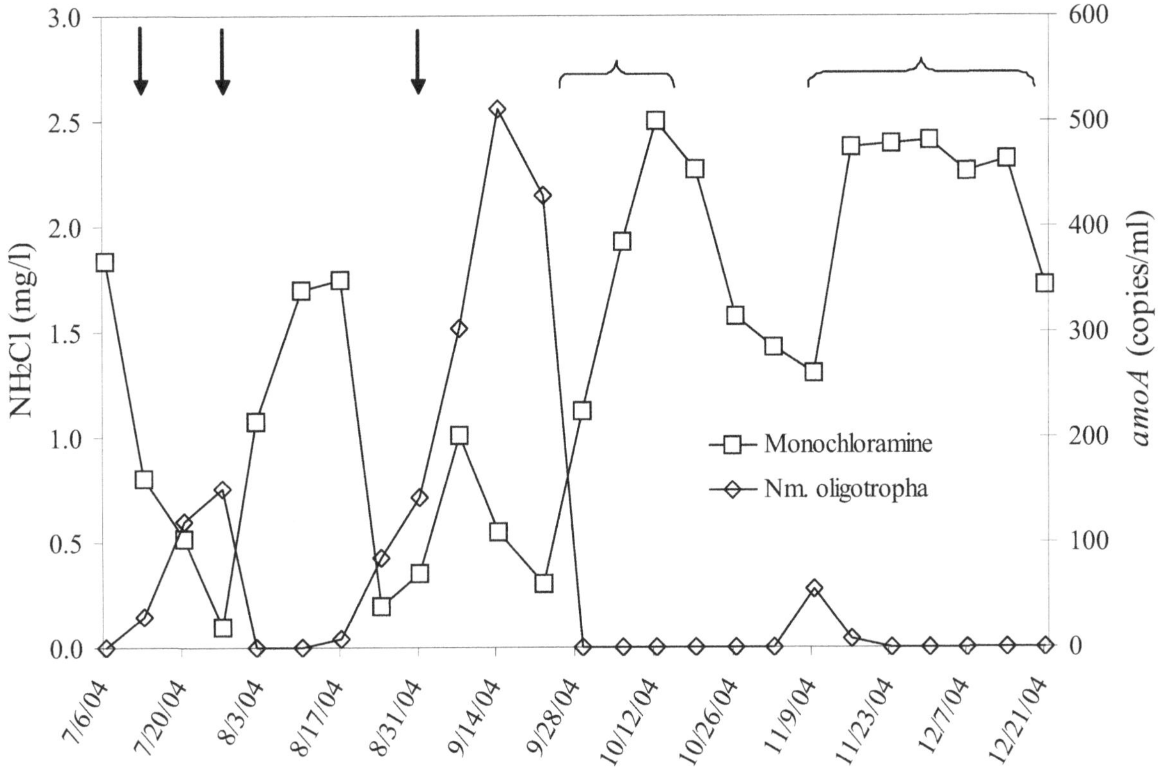

Figure 5.12 Monochloramine and real-time PCR data from Port 2 of ACWD Reservoir 8. Arrows denote breakpoint chlorination events; brackets denote booster chloramination periods.

insufficient data from this study to assess a critical AOB concentration, but this real-time PCR enumeration of AOB showed that the community was growing. The weekly sampling frequency precludes the determination of whether nitrite or AOB concentration increased concurrently or one preceded the other.

The second event that showed a change in the chemical water quality that was not reflected in the AOB concentration was on October 6th, on which the nitrite concentration increased to 26 μg/l but the AOB remained below the detection limit. A re-examination of both the water quality and operations field logs around this period do not help explain the data. On October 4th nitrite was near the detection limit at all sample locations. On October 8th the nitrite was approximately the same as the October 6th values, and on October 12th the nitrite returned to the October 4th levels. On October 14th the nitrite increased slightly (to 16 ug/L) and then fell again to the detection limit on October 19th. One explanation for these fluctuations may simply be that the nitrite data is not representing the microbial ecology of the reservoir, but rather residual nitrite from the distribution system confounding the data as system water fills the reservoir to reestablish its operating levels. We do not have system nitrite data to test this theory, and feel somewhat skeptical ourselves since additional data collected during the project indicates that the reservoir is well mixed and it would probably take a significant amount of distribution system nitrite to influence the reservoir nitrite levels. During the period in question reservoir levels were fairly constant. This event occurred during the operation of the booster chloramination system, and it is clear from the subsequent month of data that the system did not experience other symptoms of nitrification.

AOB enumeration has traditionally involved the most-probable number (MPN) assay (Soriano, and Walker 1968; Wolfe et al. 1988; Wolfe et al. 1990; Cunliffe 1991). Due to the slow growth of AOB, this assay requires a three- to four-week incubation period, ruling out its routine use in system monitoring. Apart from this delay, another concern associated with MPN enumeration is potential inaccuracies due to the inability to quantitatively recover AOB through this culture-dependant assay. A comparison of real-time PCR and MPN enumeration data, shown in Figure 5.13, indicates that the MPN protocol appears to considerably underestimate the AOB concentration in the ACWD reservoir. [Note that the *amoA* target of the real-time PCR assay is present in two or three copies on the AOB genome (Klotz, and Norton 1998).] This disparity between the two protocols might be so pronounced because of the AOB speciation in this system. The Soriano-Walker medium used in this study might have poor recovery efficiencies for *Nm. oligotropha*, which was the exclusive species detected during this study and have only been isolated using dilute ammonia concentrations (Koops et al. 1991; Suwa, Sumino, and Noto 1997). To evaluate the possibility that the real-time PCR protocol was yielding false positives, PCR amplicons were cloned and sequenced. Analysis of these sequences showed the consistent detection of *amoA* fragments derived from *Nm. oligotropha*-like AOB. Another factor that may have contributed to the differences between the results is that the real-time PCR assay does not differentiate viable and nonviable cells, provided the genomic DNA is intact following the loss of viability.

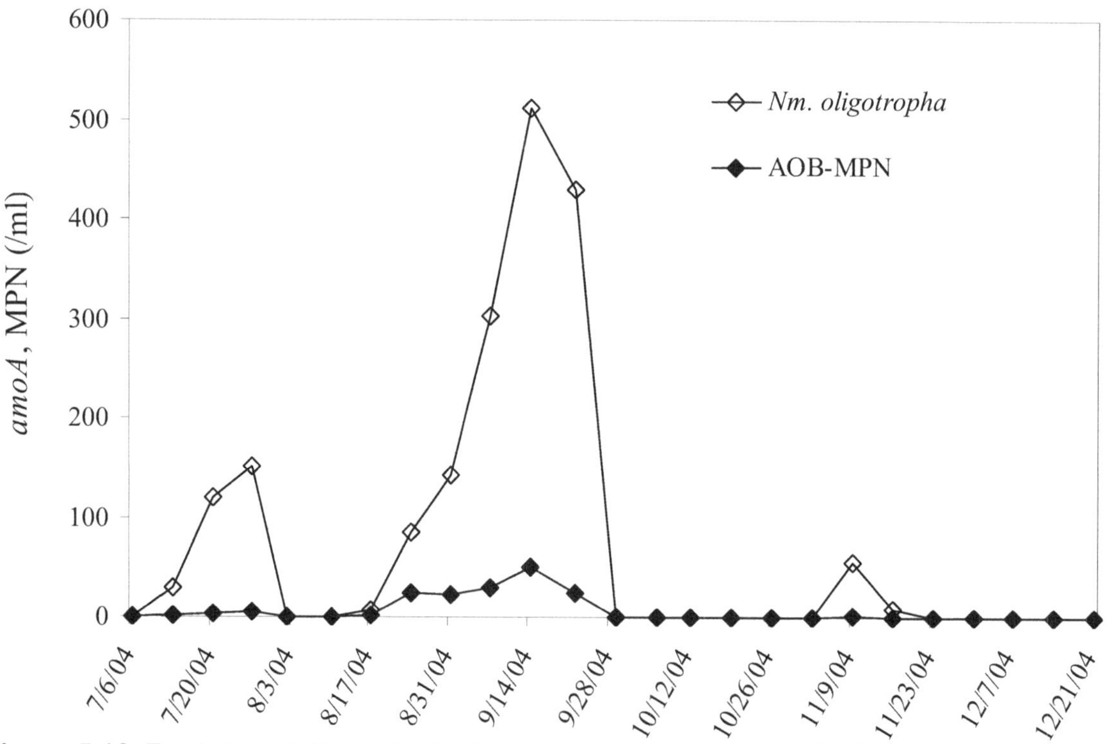

Figure 5.13 Real-time PCR and MPN data from Port 2 of ACWD Reservoir 8

Louisville Water Company Results

Water samples were collected one time from three locations in Louisville distribution system. AOB were detected in one location, Martin Hill (Table 5.15). *Nm. oligotropha*-like AOB were predominant, as observed in ACWD. In the Martin Hill sample, the monochloramine level was very low, but the concentration of nitrites was high enough to expect a nitrification episode. The *amoA* copy numbers estimated by real-time PCR were 6.75×10^3 copies, and the AOB numbers were 230 MPN/100 ml. As indicated previously, real-time PCR assay is considerably more sensitive than MPN enumeration.

Additional Studies

San Francisco Public Utilities Commission Metershop Tank Results

A closed-loop tank in the SFPUC system was sampled for several weeks during the summer of 2005, both at the tank inlet and interior. Following a period of chlorination to disinfect the tank, the feed was reverted to chloramine. The water chemistry at the tank inlet showed definite indications of nitrification, with a dramatic loss of chloramine and production of nitrite in the first week of operation with chloramine and then again after a stable 10-day period (Figure 5.14). However, the AOB data showed no detections over this entire period. There are several possible explanations for this discrepancy between chemical and microbial data. The problem may have been associated with biofilm growth and not represented in the grab sample analyses. Alternatively, nitrite may have been formed upstream of the metershop and transported to the metershop tank. If this were the case, it would be plausible to observe nitrite in the absence of AOB. It is also possible that the PCR method or sensitivity did not allow the detection of AOB in these samples. The tank interior samples showed a drop in chloramine associated with the drop in the inlet water. However, while the inlet chloramine concentration was corrected and maintained near 2 mg/L, the interior concentration continued to drop. This was not associated with a concurrent detectable production of nitrite in the tank until nearly two weeks, after which nitrite concentrations exceeded 0.25 mg N/L. AOB were not detected in the tank by real-time PCR until August 8 and 11, at levels of 23 and 29 *amoA* copies/mL, respectively. These results are in stark contrast to those from the ACWD reservoir, which showed a good correlation between suspended AOB numbers and water chemistry changes. If this discrepancy is indeed explained by biofilm growth serving as the principle source of AOB activity in this closed-loop tank system, the failure to detect AOB at all in the inlet and only at the end of the monitoring period in the tank may indicate that suspended AOB enumeration will not be successful as an early warning indicator of nitrification in systems in such systems.

Table 5.15
Chemical and biological data from LWC

Location	pH	Monochloramine ($mgCl_2$/L)	Nitrite (mgN/L)	NH3 (mgN/L)	*Nm.oligotropha* (copy numbers)	MPN (index/100ml)
Brook Hill	8.29	1.7	0-0.15	0.48	<105	<20
Martin Hill	7.95	0.3	0.3-1	0.12	6752	230
Winn Dixie	8.39	2.1	0	0.59	<105	<20

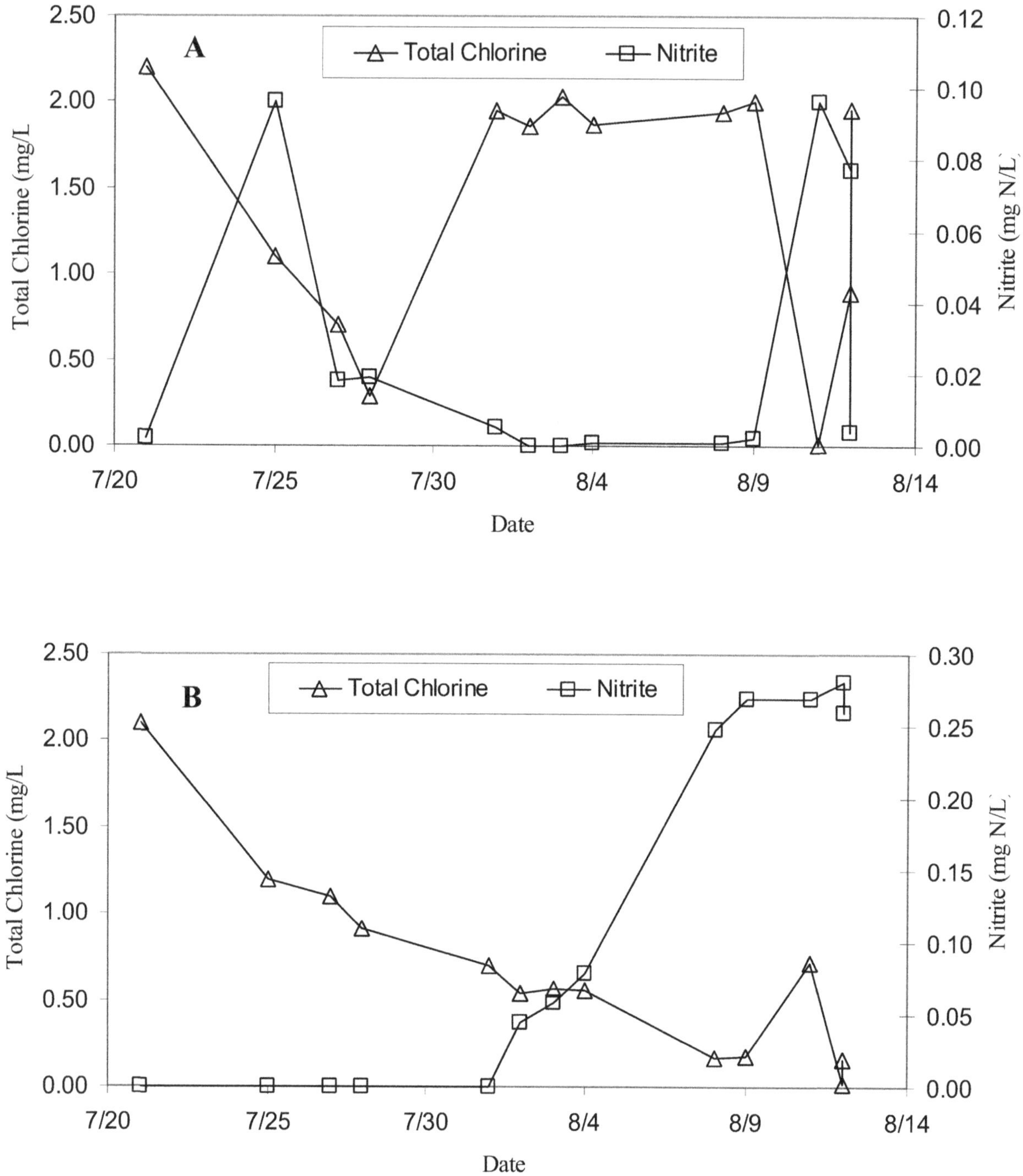

Figure 5.14 Chemical monitoring data from the Metershop Tank in the SFPUC system, (A) tank inlet, and (B) internal.

DC Water and Sewer Authority Results

A school in the DCWASA system was monitored in this study for approximately a two-month period during summer 2005, in conjunction with a recurrent warm-weather nitrification problem. Monitoring included a hydrant in the water main just outside the school building and a tap in the school kitchen. Historic chloramine and nitrite monitoring data at the hydrant and kitchen tap are shown in Figure 5.15A. With the exception of a single sampling event in November 2004, the water quality at the hydrant showed a consistent ability to maintain a high chloramine residual of between 3 and 4 mg/L, and nitrite concentrations were generally below 0.005 mg N/L (high of 0.013 mg N/L). Monitoring at the kitchen tap always revealed a drop in chloramine residual from the hydrant measurements, and there were periods of considerable chloramine loss accompanied by nitrite production at the end of 2004 and during summer 2005.

In the monitoring associated with this study (Figure 5.15B), AOB were detected only once at the hydrant (June 28) at a concentration of 10 *amoA* copies/mL, which is just above the level of quantitation. However, the closed-loop system in the school experienced significant problems in maintaining chloramine, with elevated nitrite concentrations (as high as 0.73 mg N/L) and the associated growth of AOB measured by real-time PCR. It appears clear from these data that the nitrification problem was localized to the school pipe loop, and was not originating in the water main. During the course of monitoring, there were several flushing events that consisted of leaving the kitchen sink tap open overnight, and once for a four-day period in July. The response of the system to flushing showed a very short-term improvement in water quality, with increases in chloramine and decreases in nitrite and AOB as a portion of the loop received non-impacted water from the main. An exception to this is the results following the June 21 flushing, which should the opposite trend in each parameter. An explanation of these anomalous findings is not apparent from the data. The rapid return of nitrification symptoms following flushing may indicate a role of biofilm-associated AOB that were not removed by the flushing, nor detected by the analysis of grab samples. Another possibility is that a portion of the school loop was not purged by flushing through the kitchen tap. Interestingly, during the four-day flushing in mid-July, the AOB concentration dropped below the detection limit, but the nitrite concentration only dropped to 0.4 mg N/L and the chloramine only approached 1 mg/L. This may show the influence of biofilm AOB activity in the absence of detectable AOB in suspension.

One objective of this project was to evaluate whether AOB enumeration through rapid molecular techniques could provide an early warning indication of an impending nitrification problem. The results from the DCWASA monitoring do not allow an evaluation of this hypothesis, since the monitoring began after nitrification had already started and the system never returned to water quality conditions indicative of a non-impacted system. Like the Metershop Tank results, it appears that grab sampling from the kitchen tap did not capture the real complexity of the problem, perhaps due to biofilm growth not specifically assayed in this study.

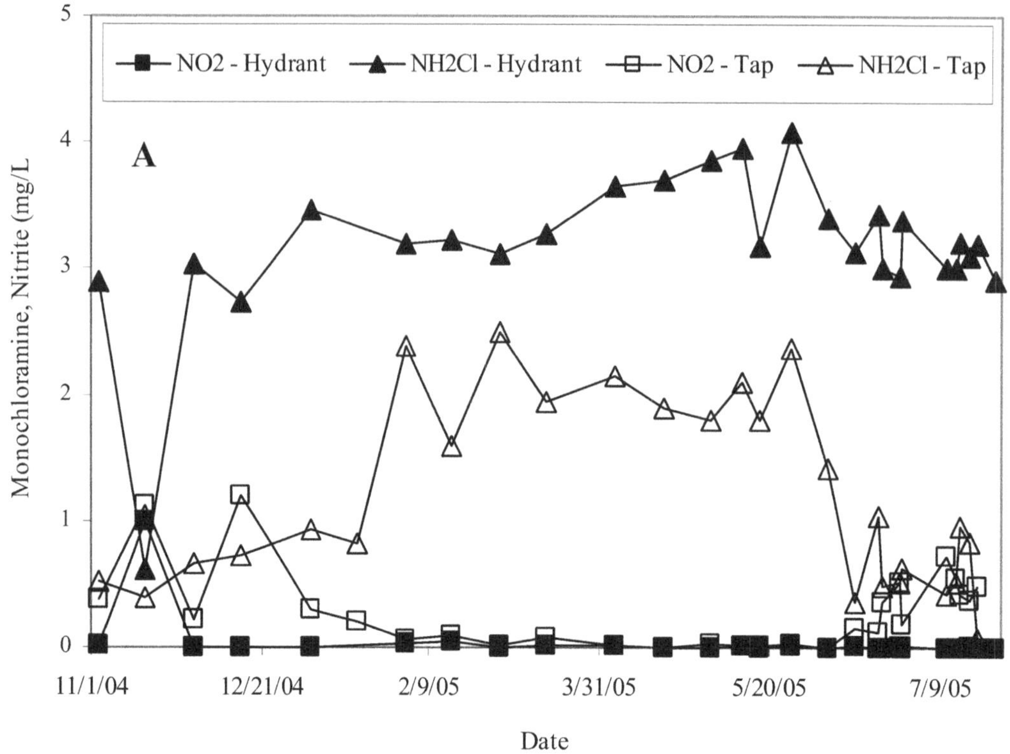

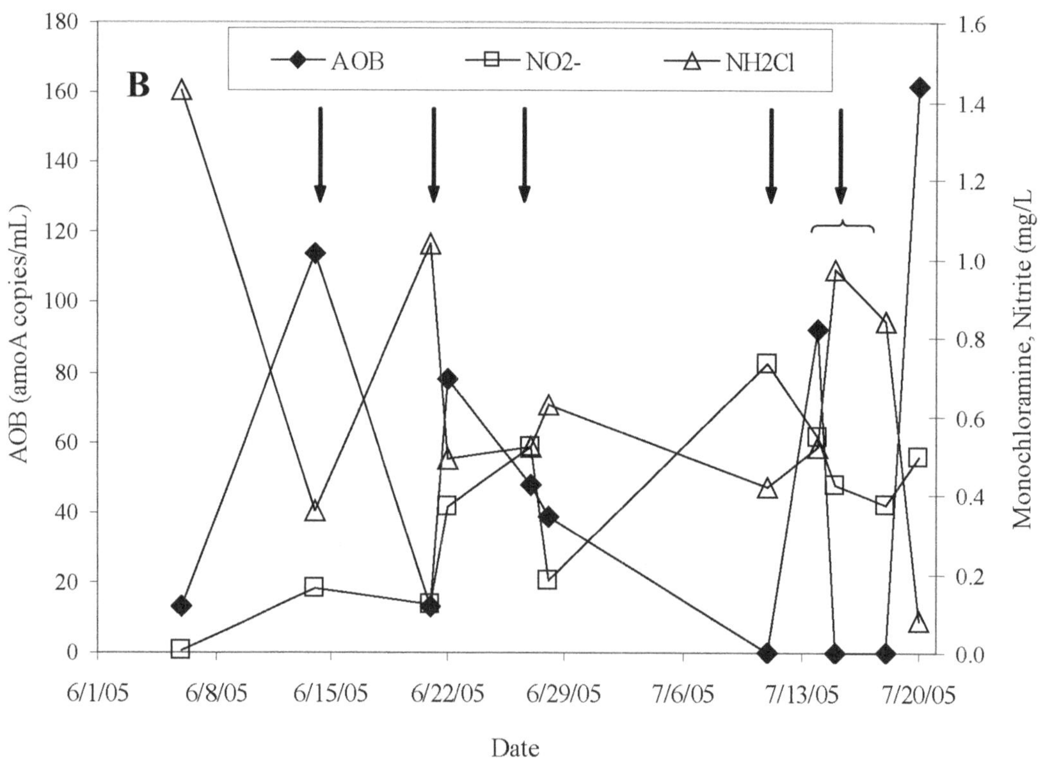

Figure 5.15 Monitoring data from the DCWASA closed-loop school system, (A) long-term chemical monitoring at the hydrant and kitchen tap, and (B) chemical and biological monitoring in this study from the kitchen tap. Flushing events are denoted with arrows.

CHAPTER 6
CONCLUSIONS AND RECOMMENDATIONS

CONCLUSIONS

The most significant conclusions derived from the results of this study are summarized as follows:

1. In FISH experiments with pure cultures, dual labeling with probes Nso1225 and AOB770 successfully increased AOB cell fluorescence relative to single-probe results.
2. TSA was a successful strategy for achieving a significant increase in the fluorescent signal of target AOB cells.
3. The FCM-based FISH protocol developed in this study exhibited the apparent loss of AOB during successive filtration and resuspension steps, with recoveries of 60-82% for *Nm. europaea* and 7-14% for *Ns. multiformis*. The compartmentalized morphology of *Ns. multiformis* may have affected the accuracy of microscopic cell counts and contributed to the apparently low cell recoveries for this species.
4. The *amoA*-targeted real-time PCR assay for AOB quantification provided excellent sensitivity and linearity, allowing quantitation down to just over 100 *amoA* copies. With the filtration of 1-L samples and correction for dilution of DNA in the PCR mixture, this offered detection limits between 4 to 6 *amoA* copies/mL. This detection limit could be lowered by filtering larger sample volumes. Based on recoveries from spiked samples, the assay did not show appreciable PCR inhibition from any of the water matrices.
5. The results from a reservoir sampled in this study showed a good correspondence between AOB numbers (determined by real-time PCR) and changes in chemical indicators of nitrification. The nitrite and AOB data generally followed the same trend, with elevated nitrite concentrations coinciding with detectable AOB concentrations relative to a non-detect background. A reduction in chloramine concentration also generally matched the AOB and nitrite changes, and no AOB were detected at chloramine concentrations above 2 mg/L in this system.
6. In two closed-loop piping systems monitored in this study, there was an apparent disconnect between suspended AOB numbers and changes in water quality parameters impacted by nitrification. This may indicate that the nitrifying community in these systems included a significant biofilm component not represented in the grab samples apart from detached cells.
7. Based on the collective findings from the systems that showed evidence of nitrification, AOB numbers from liquid grab samples do not serve as an early indication of a nitrification episode. While the reservoir system generally showed concurrent changes in biological and chemical water quality, the two pipe loop systems did not consistently demonstrate this trend.
8. The real-time PCR protocol showed a significant improvement over the MPN cultivation approach. Not only did the method yield results within one day compared to a three to four week incubation period for MPN, but it also showed dramatically higher AOB numbers.
9. Flushing of one of the pipe-loop systems in response to loss of chloramine residual and increase in nitrite was not an effective remedy for nitrification control. Evidence of nitrification returned each time in less than one week. This may be related to a biofilm-influenced AOB community.

10. Based on the several-month period during which it was monitored in this study, booster chloramination in a storage reservoir appeared to be an effective nitrification control strategy. Maintaining the chloramine level above 2 mg/L prevented AOB growth, and only during the period in which the booster system was turned off did chloramine levels drop and nitrite and AOB concentrations increase.
11. Real-time PCR results, confirmed by cloning and sequencing of *amoA* fragments, showed the exclusive detection of *Nm. oligotropha*-like AOB in the chloraminated systems monitored in this study.

RECOMMENDATIONS

The following recommendations are made based on the results of this study:

1. The use of FCM for rapid quantitation of AOB would require protocol improvements to address AOB losses associated with cell filtration and resuspension, as well as modifications to reduce the volume of reagents and the associated protocol costs. TSA-FISH was effective at boosting fluorescent signals of AOB, and coupled with FCM might provide a rapid analysis method. This might also present an effective strategy for the molecular enumeration of other microbial groups in drinking water.
2. The real-time PCR protocol used in this study presents a marked improvement over MPN enumeration, with respect to both turnaround time and cell recovery. Systems that are presently complementing their chemical analyses with MPN data might consider adapting this molecular approach.
3. The results of the pipe-loop systems, in contrast to those from a reservoir that experienced nitrification during this study, may suggest a significant role of biofilm-associated AOB in these episodes. In one of the systems, repeated flushing did not remedy the nitrification problem. In the other system, there were clear chemical indications of nitrification, but AOB were below the detection limit for over two weeks. The methods used in this study focused solely on liquid grab samples, thereby missing this biofilm contribution. Future research should investigate the role of biofilms at harboring AOB, and directly assess the effectiveness of nitrification control strategies at inactivating AOB in biofilms.
4. Booster chloramination may be an effective approach for controlling nitrification in reservoirs. While this study only monitored booster chloramination for a few months, it appeared to eradicate AOB from suspension and eliminated the associated production of nitrite.

APPENDIX

PROTOCOL FOR MICROSCOPY-BASED FISH

1. Fix samples:
   a. For cells/samples that pellet well:
      i. Add 300 μL sample to 900 μL of 4% paraformaldehyde for 30 min
      ii. Spin down cells (2 min at 13,200 rpm), wash pellet with 300 μL of 1x PBS, spin down, and resuspend in 300 μL of 1x PBS
      iii. Spot 3 μL of fixed, washed cells on each well of a 10-well Teflon-coated slide (Erie Scientific) and air dry
   b. For cells/samples that do not pellet:
      i. Add 5 μL sample on well of Teflon-coated slide and air dry
      ii. Add 15 μL paraformaldehyde for 30 min
      iii. Rinse with 2 mL of PBS
2. Dehydrate cells by immersing slides in 50, 80, and 95% ethanol gradient for 2 min each, then air dry
3. Make moisture chambers out of 50 mL polypropylene centrifuge tubes with paper towel saturated in 0.9 M NaCl. Preincubate at 46°C

   [Remaining steps performed in dark room.]

4. Add 16 μL of hybridization buffer and probe (50 ng) to each well, place slide in preincubated moisture chamber, and incubate for 1.5 hr at 46°C
5. Prepare wash buffer, and preincubate at 46°C
6. Rinse slides with 2 mL wash buffer, then immerse slides in wash buffer and incubate for 20 min at 48°C
7. Rinse slides with 2 mL H_2O, then air dry (approx 3hr)
8. Add 15 μL DAPI solution (0.5 μg/mL) to each well
9. After 5 min, wash each slide with 2 mL DW
10. Mount slides with DABCO antifadent, seal coverslip with nail polish

Standard controls: no probe as a negative control for autofluorescence, nonEUB338 as a negative control for nonspecific binding, and EUB338 as a positive control.

PROTOCOL FOR FLOW CYTOMETRY-BASED FISH

Pellet Method

1. Fix cells using steps 1(a)(i) – 1(a)(ii) above
2. Mix resuspended cells with ethanol (1:1) and store at -20°C until further processing
3. Combine 50 μL of fixed cells, 400 μL hybridization buffer, and 20 μL probe (from 50 ng/μL stock) and incubate for 3 hrs at 46°C
4. Pellet cells (3 min, 13,200 rpm)
5. Wash twice in 800 μL wash buffer for 20 min at 48°C
6. Pellet cells, resuspend in at least 500 μL cold wash buffer

Filter Method

1. Filter 100 - 1000 mL sample using 0.22 µm polycarbonate filter
2. Remove and fold the filter, transfer to a 2 mL tube containing 1 mL of TE buffer, and vortex for 2 min
3. Remove filter and check with DAPI to confirm cell detachment
4. Mix 500 µL sample with 1500 µL of 4% paraformaldehyde solution, incubate for 30 min
5. Filter sample to remove fixative, and wash filter two or three times with 2 mL PBS
6. Remove and fold the filter, transfer to a 2 mL tube containing 500 µL PBS, vortex
7. Remove the filter, mix resuspended sample with ethanol (1:1), and store at -20ºC until further processing
8. Combine 50 µL of fixed cells, 400 µL hybridization buffer, and 20 µL probe (from 50 ng/µL stock) and incubate for 3 hrs at 46ºC
9. Filter sample, wash several times with 2 mL wash buffer
10. Transfer filter to at least 500 µL cold wash buffer and vortex.

REFERENCES

Aakra, Å., J.B. Utåker, and I.F. Nes. 1999. RFLP of rRNA Genes and Sequencing of the 16S-23S rDNA Intergenic Spacer Region of Ammonia-Oxidizing Bacteria: a Phylogenetic Approach. *International Jour. of Systematic Bacteriology*, 49:123-130.

Altschul, S.F., W. Gish, E. Miller, E.W. Myers, and D.J. Lipman. 1990. Basic Logic Alignment Search Tool. *Jour. Molecular Biology*, 215:403-410.

Amann, R.I., B.J. Binder, R.J. Olson, S.W. Chisholm, R. Devereux, and D.A. Stahl. 1990. Combination of 16S rRNA-Targeted Oligonucleotide Probes With Flow Cytometry for Analyzing Mixed Microbial Populations. *Applied and Environmental Microbiology*, 56(6):1919-1925.

Amann, R.I., L. Krumholz, and D.A. Stahl. 1990. Fluorescent-Oligonucleotide Probing of Whole Cells for Determinative, Phylogenetic, and Environmental Studies in Microbiology. *Jour. Bacteriology*, 172(2):762-770.

APHA, AWWA, and WEF (American Public Health Association, American Water Works Association, and Water Environment Association). 1995. *Standard Methods for the Examination of Water and Wastewater*. 19th ed. Washington, D.C.: APHA.

American Type Culture Collection (www.atcc.org). #2265 for *Nitrosomonas europaea* and #438 for *Nitrosospira multiformis*.
http://www.atcc.org/common/documents/mediapdfs/2265.pdf
http://www.atcc.org/common/documents/mediapdfs/438.pdf

AWWA. 2000. Committee Report: Disinfection at Large and Medium-Size Systems. *Jour. AWWA*, 92(5):32-43.

Baerlocher, G.M., J. Mak, T. Tien, and P.M. Lansdorp. 2002. Telomere Length Measurement by Fluorescence *in situ* Hybridization and Flow Cytometry: Tips and Pitfalls. *Cytometry*, 47:89-99.

Belser, L.W., and E.L. Mays. 1982. Use of Nitrifier Activity Measurements to Estimate the Efficiency of Viable Nitrifier Counts in Soils and Sediments. *Applied and Environmental Microbiology*, 43(4):945-948.

Biegala, I.C., F. Not, D. Vaulot, and N. Simon. 2003. Quantitative Assessment of Picoeukaryotes in the Natural Environment by Using Taxon-Specific Oligonucleotide Probes in Association With Tyramide Signal Amplification-Fluorescence *in situ* Hybridization and Flow Cytometry. *Applied and Environmental Microbiology*, 69(9):5519-5529.

Bock, E., H.-P. Koops, H. Harms, and B. Ahlers. 1991. The Biochemistry of Nitrifying Organisms. In *Variations in Autotrophic Life*. Edited by J.M. Shively and L.L. Barton. San Diego, Calif.: Academic Press Inc.

Brodtmann, N.V., and P.J. Russo. 1979. The Use of Chloramine for Reduction of Trihalomethanes and Disinfection of Drinking Water. *Jour. AWWA*, 71(1):40-42.

Carlson, M., and D. Hardy. 1998. Controlling DBPs With Monochloramine. *Jour. AWWA*, 90(2):95-106.

Crooks, J.M., V.L. Snoeyink, M.D. Curry, and M.L. Reynolds. 1986. Technical Note: Biological Removal of Ammonia at Roxana, Illinois. *Jour. AWWA*, 78(5):94-95.

Cunliffe, D.A. 1991. Bacterial Nitrification in Chloraminated Water Supplies. *Applied and Environmental Microbiology*, 57(11):3399-3402.

DiGiano, F.A., R.A. Carrier, and A.M. Dietrich. 1986. Nitrification and Nitrosation on the Surface of GAC. *Jour. AWWA*, 76(8):70-74.

Felske, A., A.D.L. Akkermans, and W.M. DeVos. 1998. Quantification of 16S rRNAs in Complex Bacterial Communities by Multiple Competitive Reverse Transcription-PCR in Temperature Gradient Gel Electrophoresis Fingerprints. *Applied and Environmental Microbiology*, 64:4581-4587.

Fleming, K.K., G.W. Harrington, and D.R. Noguera. 2005. Nitrification Potential Curves: A New Strategy for Nitrification Prevention. *Jour. AWWA*, 97(8):90-99.

Fuchs, B.M., G. Wallner, W. Beisker, I. Schwippl, W. Ludwig, and R. Amann. 1998. Flow Cytometric Analysis of the In Situ Accessibility of *Eschericia coli* 16S rRNA for Fluorescently Labeled Oligonucleotide Probes. *Applied and Environmental Microbiology*, 64(12):4973-4982.

Gieseke, A., U. Purkhold, M. Wagner, R. Amann, and A. Schramm. 2001. Community Structure and Activity Dynamics of Nitrifying Bacteria in a Phosphate-Removing Biofilm. *Applied and Environmental Microbiology*, 67(3):1351-1362.

Haas, C.N. 1990. Disinfection. In *Water Quality and Treatment: A Handbook of Community Water Supplies*. 4th ed. New York: McGraw -Hill, Inc.

Harrington, G.W., D.R. Noguera, C.C. Bone, A.I. Kandou, P.S. Oldenburg, J.M. Regan, and D. VanHoven. 2003. *Ammonia From Chloramine Decay: Effects on Distribution System Nitrification*. Denver, Colo.: AwwaRF and AWWA.

Head, I.M., W.D. Hiorns, T.M. Embley, A.J. McCarthy, and J.R. Saunders. 1993. The Phylogeny of Autotrophic Ammonia-Oxidizing Bacteria as Determined by Analysis of 16S Ribosomal RNA Gene Sequences. *Jour. General Microbiology*, 139:1147-1153.

Heid, C.A., J. Stevens, K.J. Livak, and P.M. Williams. 1996. Real Time Quantitative PCR. *Genome Research*, 6:986-994.

Hermansson, A., and P.-E. Lindgren. 2001. Quantification of Ammonia-Oxidizing Bacteria in Arable Soil by Real-Time PCR. *Applied and Environmental Microbiology*, 67(2):972-976.

Higuchi, R., C. Fockler, G. Dollinger, and R. Waston. 1993. Kinetic PCR: Real Time Monitoring of DNA Amplification Reactions. *Biotechnology*, 11(9):1026-1030.

Horz, H.-P., J.-H. Rotthauwe, T. Lukow, and W. Liesack. 2000. Identification of Major Subgroups of Ammonia-Oxidizing Bacteria in Environmental Samples by T-RFLP Analysis of *amoA* PCR Products. *Jour. Microbiological Methods*, 39:197-204.

Hynes, R.K., and R. Knowles. 1983. Inhibition of Chemoautotrophic Nitrification by Sodium Chlorate and Sodium Chlorite: a Reexamination. *Applied and Environmental Microbiology*, 45(4):1178-1182.

Ike, N.R., R.L. Wolfe, and E.G. Means. 1988. Nitrifying Bacteria in a Chloraminated Drinking Water System. *Water Science and Technology*, 20(11/12):441-444.

Jaroszeski, M.J., and R.H. Heller. 1998. Flow Cytometry Protocols. In *Methods in Molecular Biology*. Vol 91. Totowa, N.J.: Humana Press.

Juretschko, S., G. Timmermann, M. Schmid, K.-H. Schleifer, A. Pommerening-Röser, H.-P. Koops, and M. Wagner. 1998. Combined Molecular and Conventional Analyses of Nitrifying Bacterium Diversity in Activated Sludge: *Nitrosococcus mobilis* and *Nitrospira*-Like Bacteria as Dominant Populations. *Applied and Environmental Microbiology*, 64(8):3042-3051.

Kirmeyer, G.J., G.W. Foust, G.L. Pierson, J.J. Simmler, and M.W. Lechevallier. 1993. *Optimizing Chloramines Treatment*. Denver, Colo.: AwwaRF and AWWA.

Kirmeyer, G.J., L.H. Odell, J. Jacangelo, A. Wilczak, and R. Wolfe. 1995. *Nitrification Occurrence and Control in Chloraminated Water Systems*. Denver, Colo.: AwwaRF and AWWA.

Klotz, M.G., and J.M. Norton. 1998. Multiple Copies of Ammonia Monooxygenase (*amo*) Operons Have Evolved Under Biased AT/GC Mutational Pressure in Ammonia-Oxidizing Autotrophic Bacteria. *FEMS Microbiology Letters*, 168(2):303-311.

Kolb, S., C. Knief, S. Stubner, and R. Conrad. 2003. Quantitative Detection of Methanotrophs in Soil by Novel *pmoA*-Targeted Real-Time PCR Assays. *Applied and Environmental Microbiology*, 69(5):2423-2429.

Koops, H.-P., B. Böttcher, U.C. Möller, A. Pommerening-Röser, and G. Stehr. 1991. Classification of Eight New Species of Ammonia-Oxidizing Bacteria: *Nitrosomonas communis* sp. nov., *Nitrosomonas ureae* sp. nov., *Nitrosomonas aestuarii* sp. nov., *Nitrosomonas* marina *sp. nov., Nitrosomonas nitrosa* sp. nov., *Nitrosomonas eutropha* sp. nov., *Nitrosomonas oligotropha* sp. nov. and *Nitrosomonas halophila* sp. nov. *Jour. General Microbiology*, 137(7):1689-1699.

Krasner, S.W., M.J. McGuire, J.G. Jacangelo, N.L. Patania, K.M. Reagan, and E.M. Aieta. 1989. The Occurrence of Disinfection By-Products in US Drinking Water. *Jour. AWWA*, 81(8):41-53.

Kreader, C.A. 1996. Relief of Amplification Inhibition in PCR With Bovine Serum Albumin or T4 Gene 32 Protein. *Applied and Environmental Microbiology*, 62(3):1102-1106.

Kreft, P., M. Umphres, J.-M. Hand, C. Tate, M.J. McGuire, and R.R. Trussell. 1985. Converting From Chlorine to Chloramines: A Case Study. *Jour. AWWA*, 77(1):38-45.

Layton, A.C., H. Dionisi, H.-W. Kuo, K.G. Robinson, V.M. Garrett, A. Meyers, and G.S. Sayler. 2005. Emergence of Competitive Dominant Ammonia-Oxidizing Bacterial Populations in a Full-Scale Wastewater Treatment Plant. *Applied and Environmental Microbiology*, 71(2):1105-1108.

Lieu, N.I., R.L. Wolfe, and E.G. Means, III. 1993. Optimizing Chloramine Disinfection for the Control of Nitrification. *Jour. AWWA*, 85(2):84-90.

Matulewich, V.A., P.F. Strom, and M.S. Finstein. 1975. Length of Incubation for Enumerating Nitrifying Bacteria Present in Various Environments. *Applied Microbiology*, 29(2):265-268.

McGuire, M.J., N.I. Lieu, and M.S. Pearthree. 1999. Using Chlorite Ion to Control Nitrification. *Jour. AWWA*, 91(10):52-61.

McTavish, H., J.A. Fuchs, and A.B. Hooper. 1993. Sequence of the Gene Coding for Ammonia Monooxygenase in *Nitrosomonas europaea*. *Jour. Bateriology*, 175:2436-2444.

Mendum, T.A., R.E. Sockett, and P.R. Hirsch. 1999. Use of Molecular and Isotopic Techniques to Monitor the Response of Autotrophic Ammonia-Oxidizing Populations of the ß Subdivision of the Class *Proteobacteria* in Arable Soils to Nitrogen Fertilizer. *Applied and Environmental Microbiology*, 65(9):4155-4162.

Mobarry, B.K., M. Wagner, V. Urbain, B.E. Rittmann, and D.A. Stahl. 1996. Phylogenetic Probes for Analyzing Abundance and Spatial Organization of Nitrifying Bacteria. *Applied and Environmental Microbiology*, 62(6):2156-2162.

Mobarry, B.K., M. Wagner, V. Urbain, B.E. Rittmann, and D.A. Stahl. 1997. Errata: Phylogenetic Probes for Analyzing Abundance and Spatial Organization of Nitrifying Bacteria. *Applied and Environmental Microbiology*, 63(2):815.

Neden, D.G., R.J. Jones, J.R. Smith, G.J. Kirmeyer, and G.W. Foust. 1992. Comparing Chlorination and Chloramination for Controlling Bacterial Regrowth. *Jour. AWWA*, 84(7):80-88.

Norman, T.S., L.L. Harms, and R.W. Looyenga. 1980. The Use of Chloramines to Prevent Trihalomethane Formation. *Jour. AWWA*, 72(3):176-180.

Norton, C.D., and M.W. LeChevallier. 1997. Chloramination: Its Effect on Distribution System Water Quality. *Jour. AWWA*, 89(7):66-77.

Not, F., N. Simon, I.C. Biegala, and D. Vaulot. 2002. Application of Fluorescent *in situ* Hybridization Coupled With Tyramide Signal Amplification (FISH-TSA) to Assess Eukaryotic Picoplankton Composition. *Aquatic Microbial Ecology*, 28(2):157-166.

Odell, L.H., G.J. Kirmeyer, J.G. Jacangelo, J.P. Marcinko, and R.L. Wolfe. 1996. Controlling Nitrification in Chloraminated Systems. *Jour. AWWA*, 88(7):86-98.

Okabe, S., and Y. Watanabe. 2000. Structure and Function of Nitrifying Biofilms as Determined by *In Situ* Hybridization and the Use of Microelectrodes. *Water Science and Technology*, 42(12):21-32.

Okano, Y., K.R. Hristova, C.M. Leutenegger, L.E. Jackson, R.G. Denison, B. Gebreyesus, D. Lebauer, and K.M. Scow. 2004. Application of Real-Time PCR to Study Effects of Ammonium on Population Size of Ammonia-Oxidizing Bacteria in Soil. *Applied and Environmental Microbiology*, 70:1008-1017.

Oldenburg, P.S., J.M. Regan, G.W. Harrington, and D.R. Noguera. 2002. Kinetics of *Nitrosomonas europaea* Inactivation by Chloramine. *Jour. AWWA*, 94(10):100-110.

Ootsubo, M., T. Shimizu, R. Tanaka, T. Sawabe, K. Tajima, M. Yoshimizu, Y. Ezura, T. Ezaki, and H. Oyaizu. 2002. Oligonucleotide Probe for Detecting *Enterobacteriaceae* by In Situ Hybridization. *Jour. Applied Microbiology*, 93(1):60-68.

Park, H.-D., J.M. Regan, and D.R. Noguera, 2002. Molecular Analysis of Nitrifying Bacterial Populations in Aerated-Anoxic Processes. *Water Science and Technology*, 46(1-2):273-280.

Pernthaler, A., J. Pernthaler, and R. Amann. 2002. Fluorescence *in situ* Hybridization and Catalyzed Reporter Deposition for the Identification of Marine Bacteria. *Applied and Environmental Microbiology*, 68(6):3094-3101.

Pommerening-Röser, A., G. Rath, and H.-P. Koops. 1996. Phylogenetic Diversity Within the Genus *Nitrosomonas*. *Systematic and Applied Microbiology*, 19:344-351.

Porter, J., and R.W. Pickup. 2000. Nucleic Acid-Based Fluorescent Probes in Microbial Ecology: Application of Flow Cytometry. *Jour. Microbiological Methods*, 42(1):75-79.

Purkhold, U., A. Pommerening-Röser, S. Juretschko, M.C. Schmid, H.-P. Koops, and M. Wagner. 2000. Phylogeny of All Recognized Species of Ammonia Oxidizers Based on Comparative 16S rRNA and *amoA* Sequence Analysis: Implications for Molecular Diversity Studies. *Applied and Environmental Microbiology*, 66(12):5368-5382.

Regan, J.M. 2001. Microbial Ecology of Nitrification in Chloraminated Drinking Water Distribution Systems. Ph.D. diss., The University of Wisconsin, Madison, Wis.

Regan, J.M., G.W. Harrington, H. Baribeau, R.D. Leon, and D.R. Noguera. 2003. Diversity of Nitrifying Bacteria in Full-Scale Chloraminated Distribution Systems. *Water Research*, 37(1):197-205.

Regan, J.M., G.W. Harrington, and D.R. Noguera, 2001. Diversity of Ammonia- and Nitrite-Oxidizing Bacteria in Chloraminated Distribution Systems. In *Proc. of the AWWA Water Quality Technology Conference*. Denver, Colo.: AWWA.

Regan, J.M., G.W. Harrington, and D.R. Noguera. 2002. Ammonia- and Nitrite-Oxidizing Bacterial Communities in a Pilot-Scale Chloraminated Drinking Water Distribution System. *Applied and Environmental Microbiology*, 68(1):73-81.

Rittmann, B.E., J.M. Regan, and D.A. Stahl. 1994. Nitrification As a Source of Soluble Organic Substrate in Biological Treatment. *Water Science and Technology*, 30(6):1-8.

Rotthauwe, J.-H., K.-P. Witzel, and W. Liesack. 1997. Thc Ammonia Monooxygenase Structural Gene *amoA* as a Functional Marker: Molecular Fine-Scale Analysis of Natural Ammonia-Oxidizing Populations. *Applied and Environmental Microbiology*, 63(12):4704-4712.

Ruimy, R., V. Breittmayer, V. Biovin, and R. Christen. 1994. Assessment of the State of Activity of Individual Bacterial Cells by Hybridisation With a Ribosomal RNA-Targeted Fluorescently Labelled Oligonucleotide Probe. *FEMS Microbiology Ecology*, 15:207-213.

Sambrook, J., E.F. Fritsch, and T. Maniatis. 1989. *Molecular Cloning: a Laboratory Manual*. 2nd ed. Cold Spring Harbor, N.Y.: Cold Spring Harbor Laboratory Press.

Sayavedra-Soto, L.A., N.G. Hommes, and D.J. Arp. 1994. Characterization of the Gene Encoding Hydroxylamine Oxidoreductase in *Nitrosomonas europaea*. *Jour. Bateriology*, 176:504-510.

Schramm, A., D. deBeer, M. Wagner, and R. Amann. 1998. Identification and Activities In Situ of *Nitrosospira* and *Nitrospira* spp. as Dominant Populations in a Nitrifying Fluidized Bed Reactor. *Applied and Environmental Microbiology*, 64(9):3480-3485.

Schramm, A., L.H. Larsen, N.P. Revsbech, N.B. Ramsing, R. Amann, and K.-H. Schleifer. 1996. Structure and Function of a Nitrifying Biofilm as Determined by In Situ Hybridization and the Use of Microelectrodes. *Applied and Environmental Microbiology*, 62(12):4641-4647.

Schramm, A., C.M. Santegoeds, H.K. Nielsen, H. Ploug, M. Wagner, M. Pribyl, J. Wanner, R. Amann, and D. deBeer. 1999. On the Occurrence of Anoxic Microniches, Denitrification, and Sulfate Reduction in Aerated Activated Sludge. *Applied and Environmental Microbiology*, 65(9):4189-4196.

Seidel, C.J., M.J. McGuire, R.S. Summers, and S. Via. 2005. Have Utilities Switched to Chloramines? *Jour. AWWA*, 97(10):87-97.

Sekar, R., A. Pernthaler, J. Pernthaler, F. Warnecke, T. Posch, and R. Amann. 2003. An Improved Protocol for Quantification of Freshwater *Actinobacteria* by Fluorescence *In Situ* Hybridization. *Applied and Environmental Microbiology*, 69(5):2928-2935.

Simpson, K.L., and K.P. Hayes. 1998. Drinking Water Disinfection Byproducts: An Australian Perspective. *Water Research*, 32(5):1522-1528.

Sinigalliano, C.D., D.N. Kuhn, and R.D. Jones. 1995. Amplification of the *amoA* Gene From Diverse Species of Ammonia-Oxidizing Bacteria and From an Indigenous Bacterial Population From Seawater. *Applied and Environmental Microbiology*, 61(7):2702-2706.

Skadsen, J. 1993. Nitrification in a Distribution System. *Jour. AWWA*, 85(7):95-103.

Skadsen, J. 2002. Effectiveness of High pH in Controlling Nitrification. *Jour. AWWA*, 94(7):73-83.

Smith, C.D., 2001. Chloramine Monitoring - What, Where, When. Presented at the AWWA Annual Conference and Exposition, Washington D.C.

Smith, C.D., and J.F. Smith, 1996. Excessive Loss of Chloramine Residual: A Case Study. Presented at the AWWA Annual Conference and Exposition, Toronto, Ont.

Snoeyink, V.L., and D. Jenkins. 1980. *Water Chemistry*. New York: John Wiley & Sons, Inc.

Soriano, S., and N. Walker. 1968. Isolation of Ammonia-Oxidizing Autotrophic Bacteria. *Jour. Applied Bacteriology*, 31(4):493-497.

Stehr, G., B. Böttcher, P. Dittberner, G. Rath, and H.-P. Koops. 1995. The Ammonia-Oxidizing Nitrifying Population of the River Elbe Estuary. *FEMS Microbiology Ecology*, 17:177-186.

Stephen, J.R., A.E. McCaig, Z. Smith, J.I. Prosser, and T.M. Embley. 1996. Molecular Diversity of Soil and Marine 16S rRNA Gene Sequences Related to ß-Subgroup Ammonia-Oxidizing Bacteria. *Applied and Environmental Microbiology*, 62(11):4147-4154.

Suwa, Y., T. Sumino, and K. Noto. 1997. Phylogenetic Relationships of Activated Sludge Isolates of Ammonia Oxidizers With Different Sensitivities to Ammonium Sulfate. *Jour. General Applied Microbiology*, 43:373-379.

Sykes, P.J., S.H. Neoh, M.J. Brisco, E. Hughes, J. Condon, and A.A. Morley. 1992. Quantitation of Targets for PCR by Use of Limiting Dilution. *Biotechniques*, 13:444-449.

Teske, A., E. Alm, J.M. Regan, S. Toze, B.E. Rittmann, and D.A. Stahl. 1994. Evolutionary Relationships Among Ammonia- and Nitrite-oxidizing Bacteria. *Jour. Bacteriology*, 176(21):6623-6630.

Thompson, J.D., T.J. Gibson, F. Plewniak, F. Jeanmougin, and D.G. Higgins. 1997. The ClustalX Windows Interface: Flexible Strategies for Multiple Sequence Alignment Aided by Quality Analysis Tools. *Nucleic Acids Research*, 24:4876-4882.

U.S. Environmental Protection Agency. 1998. Disinfectants and Disinfection Byproducts, Final Rule. *Fed. Reg.*, 63(241):69390-69476.

Vesey, G., D. Deere, M.R. Gauci, K.R. Griffiths, K.L. Williams, and D.A. Veal. 1997. Evaluation of Fluorochromes and Excitation Sources for Immunofluorescence in Water Samples. *Cytometry*, 29(2):147-154.

Völsch, A., W.F. Nader, H.K. Geiss, G. Nebe, and C. Birr. 1990. Detection and Analysis of Two Serotypes of Ammonia-Oxidizing Bacteria in Sewage Plants by Flow Cytometry. *Applied and Environmental Microbiology*, 56(8):2430-2435.

Wagner, M., G. Rath, R. Amann, H.-P. Koops, and K.-H. Schleifer. 1995. *In Situ* Identification of Ammonia-oxidizing Bacteria. *Systematic and Applied Microbiology*, 18(2):251-264.

Wagner, M., G. Rath, H.-P. Koops, J. Flood, and R. Amann. 1996. *In situ* Analysis of Nitrifying Bacteria in Sewage Treatment Plants. *Water Science and Technology*, 34(1-2):237-244.

Wallner, G., R. Erhart, and R. Amann. 1995. Flow Cytometric Analysis of Activated Sludge With rRNA-Targeted Probes. *Applied and Environmental Microbiology*, 61(5):1859-1866.

Watson, S.W., E. Bock, H. Harms, H.-P. Koops, and A.B. Hooper. 1989. Ammonia-Oxidizing Bacteria. In *Bergey's Manual of Systematic Bacteriology*. Vol III. Edited by J.T. Staley, et al. Baltimore, Md.: The Williams and Wilkins Co.

White, G.C. 1999. *Handbook of Chlorination and Alternative Disinfectants*. 4th ed. New York: John Wiley & Sons, Inc.

Wilczak, A., A. Assadi-Rad, H.H. Lai, L.L. Hoover, J.F. Smith, R. Berger, F. Rodigark, J.W. Beland, L.J. Lazzelle, E.R. Kincannon, J. Baker, and C.T. Heaney. 2003. Formation of NDMA in Chloraminated Water. *Jour. AWWA*, 95(9):94-106.

Wilczak, A., J.G. Jacangelo, J.P. Marcinko, L.H. Odell, G.J. Kirmeyer, and R.L. Wolfe. 1996. Occurrence of Nitrification in Chloraminated Distribution Systems. *Jour. AWWA*, 88(7):74-85.

Wintzingerode, F.v., U.B. Göbel, and E. Stackebrandt. 1997. Determination of Microbial Diversity in Environmental Samples: Pitfalls of PCR-Based rRNA Analysis. *FEMS Microbiology Reviews*, 21:213-229.

Wolfe, R.L., N.I. Lieu, G. Izaguirre, and E.G. Means. 1990. Ammonia-Oxidizing Bacteria in a Chloraminated Distribution System: Seasonal Occurrence, Distribution, and Disinfection Resistance. *Applied and Environmental Microbiology*, 56(2):451-462.

Wolfe, R.L., E.G. Means, III, M.K. Davis, and S.E. Barrett. 1988. Biological Nitrification in Covered Reservoirs Containing Chloraminated Water. *Jour. AWWA*, 80(9):109-114.

Wood, P.M. 1986. Nitrification As a Bacterial Energy Source. In *Nitrification*. Vol 20. Edited by J.I. Prosser. Oxford: IRL Press Limited.

ABBREVIATIONS

AAWU	Ann Arbor Water Utilities
ACWD	Alameda County Water District
amoA	ammonia monooxygenase A gene
AOB	ammonia-oxidizing bacteria
ATCC	American Type Culture Collection
AWWA	American Water Works Association
AwwaRF	Awwa Research Foundation
bp	basepair
BSA	bovine serum albumin
CARD	catalyzed reporter deposition
Ct	cycle threshold
DABCO	1,4-diazobicyclo[2.2.2]octane
DAPI	4',6-diamidino-2-phenylindole
DBPs	disinfection byproducts
DCWASA	DC Water and Sewer Authority
D/DBP	Disinfectants and Disinfection By-Products rule
DNA	deoxyribonucleic acid
dNTP	deoxyribonucleotide triphosphate
DO	dissolved oxygen
E	real-time PCR amplification efficiency
FCM	flow cytometry
FISH	fluorescent in situ hybridization
HAA	haloacetic acid
HPC	heterotrophic plate count
HRP	horseradish peroxidase
LWC	Louisville Water Company
MPN	most-probable number
NDMA	N-nitrosodimethylamine
NOB	nitrite-oxidizing bacteria
NOM	natural organic matter
PAC	Project Advisory Committee
PBS	phosphate-buffered saline
PCR	polymerase chain reaction

PCU	Pinellas County Utilities
rDNA	ribosomal deoxyribonucleic acid
rRNA	ribosomal ribonucleic acid
SFPUC	San Francisco Public Utilities Commission
TE	Tris-EDTA buffer
THM	trihalomethane
TOC	total organic carbon
TSA	tyramide signal amplification
USEPA	United States Environmental Protection Agency

IWA Publishing's authorised EU representative for General Product Safety Regulations is Diane D'Arras, 15 rue Duret, 75116 Paris, France, e-mail: safety@iwap.co.uk.

Printed and bound by CPI Group (UK) Ltd, Croydon, CR0 4YY
17/09/2026
02218798-0009